Praise for _Bird Sense_

"The attempt to get at what a bird sees, hears, feels and thinks is more than worth the effort because there are so many intriguing facts and stories that the reader learns along the way. Remarkable in [its] celebration of birds." —**_The New York Times_**

"A collective portrait of birds that is deeply stirring and inspires awe at our own species and its capacity for such intense curiosity." —**_The Wall Street Journal_**

"Author of highly respected technical work on the social and sexual behavior of animals, Birkhead is also capable of making his and others' research clear, and even inviting. His skill lies in the way he poses his questions . . . _Bird Sense_ asks another intriguing question: How do birds perceive their universe? What does the world look like to a bird?" —**_The New York Review of Books_**

"This is a book to make one whistle, both at the sensational senses of birds and at the patient curiosity and cunning of those who study them." —**John Spurling, _The New Republic_**

"The subtitle of British ornithologist Tim Birkhead's fantastic new book is 'What It's Like to Be a Bird.' In seven authoritative chapters, he delivers on the promise, covering sight, hearing, touch, taste, smell, birds' magnetic perceptions, and avian emotions." —**_BirdWatching_**

"By reclaiming forgotten work . . . and highlighting the studies of contemporary scientists, [Birkhead] opens up a wealth of new material to bird-loving readers. Fresh and inquiring, *Bird Sense* is a gateway title on the inner lives of birds." —*Booklist*

"[An] entertaining read perfect for birdwatchers and animal enthusiasts." —*Publishers Weekly*

"An entertaining book guaranteed to bring pleasure to bird-watchers that will also fascinate students contemplating a career in ecology." —*Kirkus Reviews*

"Birkhead provides fascinating information for birders and naturalists." —*Library Journal*

"It's hard to imagine being a bird, moving with graceful and effortless flight. But that is just the beginning . . . Tim Birkhead's *Bird Sense* brings alive the many other surprising ways that birds differ from humans in how they experience the world around them." —**Bridget Stutchbury, author of The Private Lives of Birds**

"Thoughtful, thoroughly researched and engagingly written throughout. By examining a bird's basic senses, Birkhead provides a real flavor of life from a bird's eye view." **New Scientist (UK)**

"This will be one of the seminal wildlife books of 2012 . . . Birkhead is the rare scientist who genuinely enthralls a non-technical audience. His new book will delight specialists and bedtime readers alike." —*BBC Wildlife*

BIRD SENSE

What It's Like to Be a Bird

Tim Birkhead

B L O O M S B U R Y

NEW YORK · LONDON · NEW DELHI · SYDNEY

For the sylph

Copyright © 2012 by Tim Birkhead
Illustrations © 2012 by Katrian van Grouw

Published by Bloomsbury USA, New York

All papers used by Bloomsbury USA are natural, recyclable products made from wood grown in well-managed forests. The manufacturing processes conform to the environmental regulations of the country of origin.

Every reasonable effort has been made to trace copyright holders of material reproduced in this book, but if any have been inadvertently overlooked the publishers would be glad to hear from them.

LIBRARY OF CONGRESS CATALOGING-IN-PUBLICATION DATA

Birkhead, T. R.
Bird sense : what it's like to be a bird / by Tim Birkhead.
p. cm.
Includes bibliographic references and index.
ISBN: 978-0-8027-7996-0 (hardback)
1. Birds—Physiology. 2. Birds—Behavior. 3. Birds—Psychology. I. Title.
QL698.B57 2012
598—DC23
2011043684

First published by Walker & Co. in 2012
This paperback edition published in 2013

Paperback ISBN: 978-1-62040-189-7

1 3 5 7 9 10 8 6 4 2

Typeset by Hewer Text UK Ltd, Edinburgh
Printed in the United States by Thomson-Shore, Inc. Dexter, Michigan

Contents

Preface

'Buggered' is how most New Zealanders describe their bird fauna, and it is. I've rarely been anywhere where birds are so thin in the air or on the ground. A mere handful of species – several of them flightless and nocturnal – have survived the ravages of introduced European predators, and now exist in tiny numbers, mainly on offshore islands.

The sun is already setting as we arrive at the lonely quayside. The faint purr of an outboard motor soon materialises into a small boat approaching from the island. Within minutes we are heading out to sea and into a blazing sunset. The mainland-island transition is magical: twenty minutes and we step out of the boat on to a wide, sweeping beach overhung with majestic pohutukawa trees.

Anxious to see our first kiwi we are out again as soon as we have eaten. The moonless night sky is splattered with stars – the southern Milky Way, so much more intense than that in the northern hemisphere. Our path takes us back down towards the shore and we are suddenly aware of the sea: phosphorescence! The tiny waves lapping the beach are glowing. 'You should swim,' Isabel says, and with no further encouragement we are all skinny-dipping, and ignited by bioluminescence we jump around like human fireworks. The effect is spellbinding: a visual extravaganza as elusive and astonishing as the aurora.

Ten minutes later we are dry and continue our kiwi quest into the adjacent woodland. With her infra-red camera, Isabel scans ahead, and there, hunched among the vegetation, is a dark, domed shape: our first kiwi. To the naked eye the bird is invisible, but on the camera screen it is a black blob, with an extraordinarily long, white bill. Unaware of us, the bird shuffles forward, foraging like a machine: touch, touch, touch. At the end of this long summer the ground is too hard for probing and coming across a cluster of crickets on the soil surface, the kiwi snaps them up as they attempt to hop, skip and jump away. Suddenly aware of us, the bird hurries off into the bush and out of sight. As we walk back to the house the darkness reverberates with the high-pitched squeals of male kiwis – 'k,wheee, k,wheee'.

Isabel Castro has been studying kiwis on this tiny island sanctuary for ten years. She is one of a handful of biologists trying to understand the bird's unique sensory world. Some thirty of the island's kiwis carry radio transmitters that Isabel and her students use to follow the birds' night-time wanderings and to pinpoint their daytime roosts. We have joined the annual catch-up to replace the transmitters whose batteries fade after a year.

In the brightness of the early morning sun we follow a transmitter's bleep through a forest of manuka trees and ponga (tree ferns) to a small swamp. Without speaking, Isabel indicates that she thinks our bird is in a dense patch of reeds and mimes to ask me if I'd like to catch it. Kneeling, I see a small opening in the reeds and with my face close to the muddy water I peer inside. With my head torch I can just make out a brown, hunched shape facing away from me. I wonder whether the bird is aware of me as kiwis are renowned for their deep day-time sleep. Judging the distance, I steady myself in the soggy ground and shoot my arm forward to grab the bird by its huge legs. I'm relieved: to have lost it in front of the research students would have been embarrassing. I gently pull the bird from its sleepy hollow, cradling its chest in my hands. It is heavy: at around two kilos the brown kiwi is the largest of the five (currently) recognised species.

It is not until you have this bird lying in your lap that you realise just how utterly bizarre it is. Lewis Carroll would have loved the kiwi – it is a zoological contradiction: more mammal than bird, with luxuriant hair-like plumage, an array of elongated whiskers and a long, very sensitive nose. I can feel its heart beating as I fumble my way through its plumage to find its minuscule wings. They are odd; each like a flattened finger with a few feathers along one side and a curious hooked nail at the tip (what does it use that for?). Most remarkable of all are the kiwi's tiny, all but useless eyes. Even if there had been one on the beach the previous night, the visual extravaganza of our bioluminescent cavorting would have been wasted on the kiwi.

What is it like being a kiwi? How does it feel plodding through the undergrowth in almost total darkness, with virtually no vision, but with a sense of smell and of touch so much more sophisticated than our own? Richard Owen, a nasty narcissist but a superb anatomist, dissected one around 1830 and seeing the kiwi's tiny eyes and huge olfactory region in its brain suggested – with little knowledge of the bird's behaviour – that it relied more on smell than sight. Skilfully linking form with function, Owen's predictions were elegantly borne out one hundred years later when behavioural tests revealed the laser-like accuracy with which kiwis pinpoint their prey below the ground. Kiwis can smell earthworms through 15 cm of soil! With such a sensitive nose, what does a kiwi experience on encountering another kiwi's droppings – which to me at least are as pungent as those of a fox? Does that aroma conjour up an image of its owner?

In his famous essay 'What is it like to be a bat?' published in 1974, the philosopher Thomas Nagel argued that we can never know what it feels like to be another creature. Feelings and consciousness are *subjective* experiences so they cannot be shared or imagined by anyone else. Nagel chose the bat because as a mammal it has many senses in common with ourselves, but at the same time possesses one sense – echolocation – that we do not have, making it impossible for us to know what it feels like.[1]

In one sense Nagel is right: we can never know *exactly* what it's like to be a bat or indeed a bird, because, he says, even if we imagine what it is like it is no more than that, imagining what it is like. Subtle and pedantic perhaps, but that's philosophers for you. Biologists take a more pragmatic approach, and that's what I'm going to do. Using technologies that extend our own senses, together with an array of imaginative behavioural tests, biologists have been remarkably good at discovering what it is like to be something else. Extending and enhancing our senses has been the secret of our success. It started in the 1600s when Robert Hooke first demonstrated his microscope at the Royal Society in London. Even the most mundane objects – such as a bird's feather – were transformed into something wondrous when seen through the lens of a microscope. In the 1940s biologists were amazed by the details revealed by the first sonograms – sound pictures – of birdsong, and even more amazed when in 2007 for the first time they were able to see – using fMRI (functional magnetic resonance imaging) scanning technology – the activity in a bird's brain in response to hearing the song of its own species.[2]

We identify more closely with birds than with any other group of animals (apart from primates and our pet dogs), because the vast majority of bird species – although not the kiwi – rely primarily on the same two senses as we do: vision and hearing. In addition, birds walk on two legs, most species are diurnal and some, like owls and puffins, have human-like faces, or at least, faces we can relate to. This similarity however, has blinded us to other aspects of birds' senses. Until recently it was assumed – with the kiwi as a quirky exception – that birds have no sense of smell, taste or touch. As we will see, nothing could be further from the truth. The other thing that has retarded our understanding of what it is like to be a bird is the fact that to comprehend their senses we have no alternative but to compare them with our own, yet it is this that so limits our ability to understand other species. We cannot see ultraviolet (UV) light, we cannot echolocate, nor can we sense the earth's magnetic field, as birds can, so imagining what it is like to possess those senses has been a challenge.

Because birds are so staggeringly diverse, the question 'What is it like to be a bird?' is rather simplistic and it would be much better to ask:

- What is it like being a swift, 'materialising at the tip of a long scream'?[3]
- What is it like for an emperor penguin diving in the inky blackness of the Antarctic seas at depths of up to 400 m?
- What is it like to be a flamingo sensing invisible rain falling hundreds of kilometres away that will provide the ephemeral wetlands essential for breeding?
- What is it like to be a male red-capped manakin in a Central American rainforest, displaying like a demented clockwork toy in front of an apparently uninterested female?
- What does it feel like copulating for a mere tenth of a second, but over one hundred times a day, like a pair of dunnocks? Does it wear them out or does it bring immense pleasure?
- What is it like being the lookout for a group of white-winged choughs, watching in the short term for predatory eagles; in the longer term for an opportunity to assume the breeder's mantle?
- What is it like, to feel a sudden urge to eat incessantly, and over a week or so become hugely obese, then fly relentlessly – pulled by some invisible force – in one direction for thousands of miles, as many tiny songbirds do twice each year?

These are the types of question I'm going to answer, and I'll do so by using the most recent research findings, but also by exploring how we arrived at our present understanding. We've known for centuries that we possess five senses: sight, touch, hearing, taste and smell; but in reality there are several others including heat, cold, gravity, pain and acceleration. What's more, each of the five senses is actually a mixture of different sub-senses. Vision, for example, includes an appreciation of brightness, colour, texture and motion.

Our predecessors' starting point for understanding the senses was the sense organs themselves – the structures responsible for collecting sensory information. The eyes and ears were obvious, but others, such as that responsible for the magnetic sense of birds, are still something of a mystery.

Early biologists recognised that the relative size of a particular sense organ was a good guide to its sensitivity and importance. Once the anatomists of the seventeenth century discovered the connections between sense organs and the brain, and later realised that sensory information was processed in different regions of the brain, it became apparent that the size of different brain regions might also reflect sensory ability. Scanning technology, together with good old-fashioned anatomy, now allows us to create 3-D images and measure with great accuracy the size of different regions of both human and bird brains. This has revealed, as Richard Owen predicted, that the visual regions (or centres, as they are known) in the kiwi's brain are almost non-existent, yet its olfactory centres are even larger than he thought.[4]

Once electricity had been discovered in the eighteenth century, physiologists like Luigi Galvani quickly realised that they could measure the amount of 'animal electricity' or nervous activity in the connections between sense organs and the brain. As the field of electrophysiology developed it became clear that this provided yet another key to understanding the sensory abilities of animals. Most recently, neurobiologists have used different types of scanners to measure activity in different regions of the brain itself to inform them of sensory abilities.

The sensory system controls behaviour: it encourages us to eat, to fight, to have sex, to care for our offspring, and so on. Without it we couldn't function. Without any one of our senses life would be so much poorer and much more difficult. We strive to feed our senses: we love music, we love art, we take risks, we fall in love, we savour the scent of freshly cut meadows, we relish tasty food and we crave our lover's touch. Our behaviour is controlled by our senses and, as a result, it is behaviour that provides one of the easiest ways of deducing the senses that animals use in their daily lives.

The study of senses – and bird senses in particular – has had a chequered history. Despite the abundance of descriptive information accumulated over the past few centuries, the sensory biology of birds has never been a hot topic. I avoided sensory biology as a zoology undergraduate in the 1970s, partly because it was taught by physiologists rather than behaviourists, and partly because the links between the nervous system and behaviour were known only for what I considered rather unexciting animals like sea slugs, rather than for birds.

Part of my motivation for writing this book, then, is to make up for lost time. I have also been encouraged by a change in attitude, not so much among physiologists but among my animal behaviour colleagues, who during recent decades have effectively rediscovered the sensory systems of birds and other animals. While I was writing this book I contacted several retired sensory biologists and was surprised to discover that they all had a similar tale: *when I was doing this research no one was interested, or they didn't believe what we had found.* One researcher told me how his entire career had been devoted to the sensory biology of birds and, apart from once being asked to write a chapter for an encyclopaedia of bird biology, had received relatively little recognition. On retirement he had burned all his papers, and then – to his simultaneous dismay and delight – I started asking him about his research.

Others told me how they had once planned to write a textbook on the sensory biology of birds but failed to find a publisher sufficiently interested. I cannot imagine what it must be like devoting your life to an area of research that few others find interesting. However, different areas of biology flourish at different times and I am optimistic that the sensory biology of birds is about to have its day.

So what's changed? From my own perspective, the field of animal behaviour has changed dramatically. I describe myself as a behavioural ecologist first and an ornithologist second: a behavioural ecologist who studies birds. Behavioural ecology is a branch of animal behaviour that emerged in the 1970s, with a tight focus on the adaptive significance of behaviour. The behavioural ecologist's approach was to ask how a particular behaviour increases the chances of an individual passing on its genes to the next generation. For example, why does the buffalo weaver – an African, starling-sized bird – copulate for thirty minutes at a time when most other birds copulate for just a couple of seconds? Why does the male cock-of-the-rock display in groups of other males and play no part in rearing his offspring?

Behavioural ecology has been extraordinarily successful in making sense of behaviours that to previous generations had been a mystery. But behavioural ecology has also been a trap, for like all disciplines its boundaries have restricted researchers' horizons. As the subject matured during the 1990s many behavioural ecologists began to realise that, on its own, identifying the adaptive significance of behaviour wasn't enough. Back in the 1940s, when the study of animal behaviour was in its infancy, one of its founders, Niko Tinbergen (later a Nobel Laureate), pointed out that behaviour could be studied in four different ways: by considering its (i) adaptive significance; (ii) causes; (iii) development – how the behaviour develops as the animal grows up; and (iv) evolutionary history. By the 1990s behavioural ecologists, whose entire focus during the previous twenty years had been on the adaptive significance of behaviour, began to realise that they needed to know more about the other aspects of behaviour and, in particular, the causes of behaviour.[5]

Let's see why. The zebra finch is a popular study species for behavioural ecologists, especially for studies of mate choice. Female zebra finches have an orange beak and males a red beak, a sex difference that suggests that the male's brighter beak colour evolved because females prefer a redder beak. Some, but not all, behavioural tests suggest that this is true and researchers assume that because *we* can rank the beaks of male zebra finches from orangey-red to blood-red, female zebra finches can do the same. They have never tested this assumption in terms of what zebra finches can actually see, yet it is widely assumed that beak colour is an important component of female choice.[6]

Another trait that female birds are thought to use in their selection of a mate is the symmetry of plumage markings, such as the pale spots on the throat and chest of male European starlings. Careful tests in which female starlings were 'asked' to discriminate between different levels of plumage symmetry (using images rather than live birds) revealed that, while they could identify males that were highly asymmetric in their spotting, their ability to discriminate smaller differences was not very good. In fact, to a female starling most males look much the same in this respect, demonstrating that they were unlikely to use plumage symmetry as a way of choosing a male.[7]

Behavioural ecologists have also assumed that the degree of sexual dimorphism in birds – that is, how different males and females are in their appearance – might be linked to whether they are monogamous or polygamous. To test this they scored species according to the brightness of the male and female plumage – based on *human* vision. We know now that this is naive, for the avian visual system is not like ours because birds can see ultraviolet (UV) light. Scoring the same birds under UV light revealed that a large number of species – including the blue tit and several parrots – previously thought to lack sexual dimorphism, actually differed quite a lot when viewed – as females would see them – with UV vision.[8]

As these examples illustrate, of all avian senses, vision – and colour vision in particular – is the area where the most spectacular recent discoveries have been made, mainly because this is where researchers have focused most effort.[9] Researchers now realise that to understand the behaviour of birds it is essential to understand the kind of worlds they live in. We are just beginning to appreciate, for example, that many birds other than kiwis have a sophisticated sense of smell; that many have a magnetic sense that guides them on migration, and, most intriguingly of all, that like us birds have an emotional life.

What we know about the senses of birds has been acquired gradually over centuries. Knowledge accumulates by building on what others have previously found, and, as Isaac Newton said, by standing on the shoulders of giants. Because researchers feed on each other's ideas and discoveries, and since they both collaborate and compete with each other, the more individuals there are working on a particular topic, the more rapidly progress is made. Progress is accelerated, of course, by intellectual giants: think Darwin for biology, Einstein for physics and Newton for mathematics. But scientists are human, too, and susceptible to human foibles, and progress isn't always rapid or straightforward. It is all too easy to become fixated on one idea – as we'll see. Research is full of blind alleys and scientists constantly have to judge whether to persist in what they believe to be correct, or to give up and try a different line of enquiry.

Science is sometimes described as a search for the truth. This sounds rather pretentious, but 'the truth' here has a straightforward meaning: it is simply what, on the basis of the available scientific evidence, we currently believe. When scientists retest someone else's idea and find that evidence to be consistent with the original notion, then the idea remains. If, however, other researchers fail to replicate the original results, or if they find a better explanation for the facts, scientists can change their idea about what the truth is. Changing your mind in the light of new ideas or better evidence constitutes scientific progress. A better term, then, is 'truth for now' – on the basis of the *current* evidence this is what we believe to be true.

The evolution of the eye is a good example of how our knowledge has progressed. Throughout much of the seventeenth, eighteenth and nineteenth centuries, it was believed that God in his infinite wisdom had created all life forms and had given them eyes to see with: owls have especially large eyes because they need to see in the dark. This way of thinking about the perfect fit between an animal's attributes and its lifestyle was called 'natural theology'. But there were some things that simply didn't look like God's wisdom: why males produced so many sperm, for example, when only one was needed for fertilisation. Would a wise God be so profligate? Charles Darwin's idea of natural selection, presented in the *Origin of Species* in 1859, provided a much better explanation for all aspects of the natural world than the wisdom of God, and as the evidence accumulated scientists abandoned natural theology in favour of natural selection.

Scientific studies usually begin with observations and descriptions of what something *is*. Once again, the eye provides a good example. Starting in ancient Greece, the early anatomists removed the eyes of sheep and chickens and cut them open to see how they were constructed, and made detailed descriptions of what they saw – and sometimes what they imagined they saw. Once the descriptive phase is complete, scientists start to ask other kinds of questions, such as 'How does it work?' and 'What is its function?' Often, while one kind of biologist may be an expert in anatomy and can provide a detailed description, it usually requires a different range of skills to understand how something like the eye actually works. As our knowledge increases and researchers become more and more specialised in their knowledge, they usually have to collaborate with others whose skills complement their own. Understanding how the eye works today, for example, requires expertise in several different fields, including anatomy, neurobiology, molecular biology, physics and mathematics. It is this interdisciplinary approach – the interactions between researchers with different kinds of expertise – that ultimately makes science both exciting and successful.

Ideas have a particularly important place in science. Having an idea about why something is the way it is is crucial since it provides the framework for asking questions – and asking the *right* questions. For example, why do the eyes of owls face forward, whereas those of ducks are directed sideways? One idea for the owl's forward-facing eyes is that, like us, owls rely on binocular vision for depth perception. But there are other ideas, too, some of which, as we shall see, are even better supported by the evidence.

Ideas are important in another way because, if an idea results in a discovery, then this can make a scientist's reputation. Science is about being first, and about being the person(s) associated with a particular discovery, exemplified by James Watson's and Francis Crick's discovery of the structure of DNA in 1953.

Where, you might ask, do scientists get their ideas from? Partly from the body of knowledge they already have, partly from discussing their work with other scientists, but sometimes from casual observations or comments made by non-scientists. As we'll see, casual comments have played a vital role in alerting scientists to particular bird senses. One of the most intriguing, described later, is the sixteenth-century Portuguese missionary in Africa who recounted how, whenever he lit beeswax candles, little birds came into the sacristy to eat the melted wax.

Once a scientist has had an idea and tested it in the most rigorous way she or he can devise, usually through some kind of experiment, they might then present their results by giving a talk at a scientific conference. This allows them to gauge how others view their results. On the basis of this the scientist may or may not modify their interpretation. The next stage is to write up the results so that they can be published as a paper in a scientific journal. The editor of the journal receives the scientist's report and sends it off to two or three other scientists (referees) who decide whether or not it merits publication. Their comments in turn may provide the author with new ideas and an opportunity to re-analyse some of their results and modify the report. If, on the basis of the referees' comments, the manuscript is deemed acceptable it is then published either as a hard copy and/or an on-line version in the scientific journal. Even then, the process isn't over, for once the report is published it becomes available to all other scientists, who can either criticise it or take inspiration from it for their own studies.

In a nutshell, then, this is the tried and tested process of science and it hasn't changed much since the late 1600s, when the first scientific journal was published. Throughout this book we'll meet the individuals who, through an unequal combination of perspiration and inspiration, are responsible for the scientific discoveries associated with bird senses. Typically, the accounts of their discoveries published in scientific journals are written concisely and with a fair amount of jargon – in both cases to save space. Jargon isn't a problem for those in the same field, but to those outside a particular research area, and to the non-specialist, it can be a major obstacle to understanding. What I have done in this book is to take the scientific papers relating to the senses of birds, and convey their findings in everyday language. I have avoided jargon wherever possible, but where it is completely unavoidable I have tried to briefly explain the term, and for those requiring a bit more explanation I have provided a glossary at the back of the book. One of the benefits of writing what I hope is an accessible account of the senses of birds is that it made me ask my sensory colleagues some rather basic questions. In doing so I discovered that there were many aspects where I assumed the answers would be known, only to find that there is a great deal still to be discovered. This is inevitable since we cannot know everything, but, of course, it can be a little frustrating when we find that the answers to what seem like very simple question are not known. On the other hand, such gaps in our knowledge are exciting because they identify new opportunities for researchers interested in the senses of birds.

Bird Sense is about how birds perceive the world. It is based on a lifetime of ornithological research and a conviction that we have consistently underestimated what goes on in a bird's head. We already know quite a lot, and we are poised to make more discoveries. This is the story of how we got to where we are, and what the future holds.

My entire career has been spent studying birds. That doesn't mean I do nothing else: being an academic in a university means I also spend a fair amount of time teaching undergraduates (which I enjoy), and rather less time on administration (which I don't). I started watching birds at the age of five, encouraged by my father, and was lucky to be able to turn my passion for birds into a career as a scientist. It is a career that's taken me all over the world, from the Arctic to the tropics, studying birds. As a result, and largely by working with my research students and colleagues, I have acquired privileged insight into the biology of a fair number of different bird species. Two, however, have preoccupied me: the zebra finch and the common guillemot. My experience of keeping zebra finches and other birds as a boy, combined with endless hours watching wild birds, honed my observational skills and gave me what I like to think of as a kind of biological intuition about the way birds operate. Hard to define, but I'm sure the many hours I spent watching birds helped to make me an effective researcher. Certainly, it set me up for the twenty-five years, so far, that I have spent studying zebra finches.

My other main study species is the common guillemot/ormurre (North America). This was the subject of my PhD and I spent four blissful summers on Skomer Island, off the western tip of South Wales, studying this species' breeding behaviour and ecology. That was almost forty years ago, and I have been back to Skomer and its guillemots almost every summer since then. As I was writing this I realised that I have probably devoted more time to watching and thinking about guillemots than to any other species. This is reflected in the book, for guillemots have given me tremendous insight into what it's like to be a bird.

Probably not all scientific ornithologists feel that way about their study species, but I certainly do and I think – at the risk of seeming anthropomorphic – it is because guillemots are so similar to humans. They are extremely social, forming friendships with their neighbours and occasionally helping them with childcare; they are monogamous (albeit with the occasional fling); male and female pair members work together to rear offspring, and pairs sometimes remain together for as long as twenty years.

The other benefit of studying birds for so long is that one gets to know, either personally or by email, a large number of other ornithologists, and perhaps the most rewarding aspect of writing this book has been the enthusiasm with which my colleagues have shared their hard-won knowledge. Without exception, everyone I contacted to ask questions or to seek clarification responded helpfully. I am grateful to them all (and I apologise if I've overlooked anyone): Elizabeth Adkins-Regan, Kate Ashbrook, Clare Baker, Greg Ball, Jacques

Balthazart, Herman Berkhoudt, Michel Cabanac, John Cockrem, Jeremy Corfield, Adam Crisford, Susie Cunningham, Innes Cuthill, Marian Dawkins, Bob Dooling, Jon Erichsen, John Ewen, Zdenek Halata, Peter Hudson, Alex Kacelnik, Alex Krikelis, Stefan Leitner, Jeff Lucas, Helen Macdonald, Mike Mendl, Reinhold Necker, Gaby Nevitt, Jemima Parry-Jones (of the International Bird of Prey Centre), Larry Parsons, Tom Pizzari, Andy Radford, Uli Reyer, Claire Spottiswoode, Martin Stevens, Rod Suthers, Eric Vallet, Bernice Wenzel and Martin Wild. I am especially grateful to Isabel Castro who, having promised me a kiwi experience of a lifetime, didn't disappoint. Thanks to Geoff Hill for taking me kayaking in the swamps of Florida in search of the ivory-billed woodpecker; we didn't see one but the experience was unforgettable. Special thanks to Patricia Brekke for persuading me to visit Tiritiri Matangi Island in New Zealand to see her stitchbirds; to Claire Spottiswoode for introducing me to the wonders of honeyguides and prinias in Zambia; to Ron Moorehouse who arranged for me to visit Codfish Island, New Zealand, to see kakapo up close – an extraordinary privilege for which I am very grateful. I thank Nicky Clayton for patiently answering my questions about cognition. Peter Gallivan and Jamie Thomson provided some much-appreciated help with the references. Graham Martin kindly read and commented on chapter 1, and Herman Berkhoudt did the same for chapter 3. I am particularly grateful to Bob Montgomerie for years of constructive criticism and friendship, and for reading and commenting on the entire manuscript. Similarly, I thank Jeremy Mynott for his perceptive comments on the manuscript. My agent Felicity Bryan provided her usual invaluable advice, and Bill Swainson and his team at Bloomsbury were exemplary in their support. As always, I thank my family for their indulgence.

I

Seeing

The wedge-tailed eagle has the largest eye relative to body size of any bird. Thumbnails (*from left to right*): the retina of an eagle showing the two fovea and the pecten (*dark*); a cross section through an eagle's eye; cross section through an eagle's skull showing the relative size and position of the eyes and the line of sight of the two fovea (*arrows*).

The falcon's sensory world is as different from ours as is that of a bat or a bumble-bee. Their high-speed sensory and nervous systems give them extremely fast reactions. Their world moves about ten times faster than ours.

Helen Macdonald, 2006, *Falcon*, Reaktion Books

A s a child I once had a conversation with my mum about what our dog could or could not see. On the basis of something I had heard or read, I told her that dogs could see only in black and white. Mum was not impressed. 'How could they possibly know that?' she said: 'We cannot look through a dog's eyes, so how could anyone know?'

In fact there are several ways we can know what a dog, a bird or, indeed, any other organism can see, for example either by looking at the structure of the eye and comparing it with other species, or by behavioural tests. In the past, falconers unwittingly performed just such a test – not with falcons, but with shrikes.

> This elegant little bird is used, not to attract the hawk as might be supposed, but to give notice of its approach. Its power of vision is perfectly marvellous, for it will detect and announce the presence of a hawk in the air long before the latter is discernible by human eye.[1]

The 'elegant little bird' is the great grey shrike and the trapping method an elaborate one, involving a turf hut in which the falconer is concealed, a live decoy falcon, a wooden decoy falcon, a live pigeon and – crucially – a great grey shrike (known also as a butcher bird) tethered outside its own miniature turf hut.

James E. Harting, falconer and ornithologist, saw this method in action during October 1877 near Valkenswaard, in the Netherlands, a traditional location for trapping migrating falcons. Here's how he described it:

> We take our seats on the chairs in the hut, and fill our pipes . . . Suddenly our attention is attracted by one of the shrikes. He chatters and appears uneasy. He crouches and points . . . He jumps off the roof of his hut, and prepares to take shelter within it. The falconer says there is a hawk in the air.[2]

They watch and wait, but it turns out to be a buzzard and the falcon isn't interested. But later:

> Look! The butcher-bird is pointing again. There is something in the air. He chatters and quits his perch . . . We look in the direction indicated, and strain our eyes, but see nothing. 'You will see him presently', says the falconer; 'the butcher-bird can see much farther than we can.' And so he can. Two or three minutes afterwards on the far distant horizon of that great plain [of Valkenswaard] a speck comes into view, no bigger than a skylark. It is a falcon.[3]

As the raptor approaches, the nature of the shrike's agitation informs the falconer of the species. Even more remarkably, the shrike's behaviour also tells the falconer *how* the raptor is approaching: swiftly or slowly; high in the sky or low over the ground. The shrike – an invaluable asset – is kept safe from the raptor's clutches by the provision of that little turf hut.

Other trapping methods employed shrikes as decoys, relying on the extraordinary visual acuity of the raptors to see them as potential prey. Expressions such as 'eagle-eyed' or 'hawk-eyed' attest to the fact that for a very long time we have known about the extraordinary vision of falcons and other birds of prey.[4]

One reason falcons see so well is because they have *two* visual hot spots at the back of each eye – two foveas – rather than the one that humans have. The fovea is simply a tiny pit or depression on the retina at the back of the eye where blood vessels are absent (since they would interfere with the clarity of the image) and the density of photoreceptors – cells for detecting light – is greatest. As a result, the fovea is the point in our retina where the image is sharpest. The falcon's two foveae contribute to its excellent vision.

Around half of all bird species examined so far have a single fovea, like us, and the question is whether shrikes have one or two. When I asked my academic colleagues who specialise in avian vision, no one knew. But one told me where to look: 'Check Casey Wood's *Fundus Oculi*,' he said. Remarkably, I knew of this obscurely titled book, published in 1917, although I had never perused it. Wood's *Fundus Oculi* is a study of the retinas of birds, as viewed through an optician's ophthalmoscope. Its title – which guaranteed that it would never be a bestseller – refers simply to the back of the eye.

Casey Albert Wood (1856–1942) was already one of my heroes. Professor of ophthalmology at the University of Illinois between 1904 and 1925, and probably the most eminent eye specialist of his age, Wood was also fascinated by birds, bird books and the history of ornithology. Recognising, for example, the immense significance of Frederick II's thirteenth-century manuscript on falconry (and ornithology), Wood went to the Vatican Library, translated it and published it, making this extremely rare manuscript much more widely available. He also discovered, and purchased for his personal library, a unique hand-coloured copy of Willughby and Ray's *Ornithology* (1678) that John Ray had presented to Samuel Pepys when Pepys was president of Britain's Royal Society in the

1680s. Another major achievement was Casey Wood's *Introduction to the Literature on Vertebrate Biology*, a remarkable reference book that I own and use regularly, that lists *all* known zoology books (including those on birds) published before 1931.

Wood's *The Fundus Oculi of Birds* (to give it its full title) grew out of his belief that a better understanding of the exceptional eyesight of birds would throw light on the biology and pathology of human vision. It was a stroke of genius and, employing the same equipment he used to examine the human retina, Wood described and catalogued the eyes of a wide range of living bird species. Such was his knowledge that it was said he could identify a bird simply from an image of its retina![5]

My first opportunity to look at Wood's *Fundus Oculi* occurred during a visit to the ornithological Blacker-Wood Library at McGill University, Montreal, which I visited while searching for material for my book *The Wisdom of Birds* (2009). Casey Wood had donated his huge personal library to the university in honour of his wife. I went with my colleague Bob Montgomerie, specifically to look at the Pepys' *Ornithology*, and while I was there Eleanor MacLean, the librarian, asked if I'd also like to look at the *Fundus Oculi*. Stupidly, I declined, befuddled by its title and distracted by too many other more interesting old books.

Even if I had looked at it, there was no way I would have remembered whether Casey Wood had included shrikes in his survey, and when I later needed it I discovered that the book was scarce in British libraries. I eventually found one, and there, under 'California Shrike *Lanius ludovicianus gambeli*', now known as the loggerhead shrike, Wood writes: 'There are two macular regions in the fundus of this bird.' In other words, yes, there are two foveas (macular regions) on the back of the eye (the fundus) of the loggerhead shrike. Excellent! Just as I hoped, and as Wood says: 'Birds with double foveae have exceptionally good eyesight.'[6]

The human eye has long fascinated lovers, artists and physicians. The ancient Greeks dissected eyes, but struggled to understand how they worked, unclear as to whether they received or emanated light. The anatomical descriptions of the eye made by Galen – physician to the Roman gladiators during the second century AD – remained the standard until the Renaissance, when there was renewed interest in the natural world, and in the wonder of vision, inspired by translations of Islamic manuscripts from the thirteenth and fourteenth centuries. The German polymath Johannes Kepler (1571–1630) was among the first to create a theory of vision, later elaborated by Isaac Newton, René Descartes and many others. In 1684 Antonie von Leeuwenhoek, pioneer microscopist, got the first glimpse of what we now know to be light-sensitive cells – the so-called rods and cones – in the retina. Two hundred years later, using a much better microscope and a very clever way of staining different types of cell-different colours, Santiago Ramón y Cajal (1852–1934) provided a wonderfully detailed – and exquisitely illustrated – description of the way the cells of the retina connect to the brain, in a variety of animals, including birds.

In the *Origin of Species* Darwin refers to vertebrate eyes as 'organs of extreme perfection and complication'. In a sense, the eye was a test case for natural selection because the Christian philosopher William Paley had used the eye in his *Natural Theology* (1802) as an example of the Creator's wisdom. Only God, Paley asserted, could produce an organ so perfectly adapted for its purpose. Paley called it a 'cure for atheism'. As an undergraduate at Cambridge, Darwin had enjoyed Paley's book, when, believe it or not, he was training for the Church. But as Darwin said later, Paley's ideas about the natural world (which were essentially about adaptation) all seemed quite plausible – before his discovery of natural selection. The recognition that natural selection provided a much more convincing explanation than God or natural theology for the perfection of

the natural world was one of the fundamental shifts in our understanding of nature.

Paley was a creationist and an advocate of 'intelligent design',
and the crux of his argument was that half an eye was of no use, and
that therefore natural selection could not possibly create an eye. For
Paley and the creationists the eye had to be fully developed to be of
any use, and the only way that could happen was if God made it.

The flaw in this way of thinking has been exposed many times,
but most tellingly in an ingenious reconstruction of the evolution
of the eye by two Swedish scientists, Dan-Eric Nilsson and Susanne
Pelger, in 1994. Starting from a simple sheet of light-sensitive cells,
they showed that a 1 per cent improvement in vision each generation could generate a sophisticated eye similar to that of a human or
a bird in less than half a million years – a relatively short period in
the history of life on earth. This evolutionary model not only
showed that half an eye (or less) was better than no eye at all; it
also revealed that the evolution of vision was nowhere near as
complicated (or impossible) as Paley and his followers believed.[7]

As I read more about the eyesight of birds, one particular phrase
kept cropping up: *a wing guided by an eye*, meaning that a bird is no
more than a flying machine with excellent vision. After a while I
began to feel a twinge of irritation every time I read it, because it
implies that vision is the *only* sense birds have: but, as we will see,
nothing could be further from the truth. The expression comes
from a book on vertebrate vision published in 1943 by a French
ophthalmologist, André Rochon-Duvigneaud (1863–1952), for
whom the aphorism captured the essence of being a bird.

Of course, long before Rochon-Duvigneaud, almost everyone
who has written about birds has commented on their excellent
eyesight. The great French naturalist the Comte de Buffon, for
example, discussing the senses of birds in the 1790s, said: 'We find
that of sight to be more extended, more acute, more accurate and

more distinct in the birds in general, than in the quadrupeds' and 'A bird . . . that shoots swiftly through the air, must undoubtedly see better than one which slowly describes a waving tract' – meaning a bird with a slower, more meandering flight.[8] Then, in the early nineteenth century, the ornithologist James Rennie wrote: 'We have ourselves more than once seen the osprey dash down from a height of two or three hundred feet upon a fish of no considerable size, and which a man could with difficulty have perceived at the same distance' and 'The bottle tit [long-tailed tit] flits with great quickness among the branches of trees, and finds on the very smooth bark its particular food, where nothing is perceptible to the naked eye, though insects can be detected there by the microscope.'[9] In a similar vein, there is an oft-repeated observation that an American kestrel can detect a two-millimetre-long insect at a distance of 18 metres.[10] Unsure about what that meant in terms of human vision, I checked and, yes, at a distance of 18 m a two-millimetre-long insect is completely invisible, and in fact did not become visible until I had approached to within four metres – striking evidence of the kestrel's superior visual resolution.

While conducting my PhD on guillemots on Skomer Island I constructed hides at various colonies to be able to watch their behaviour at close range. One of my favourite hides was on the north side of the island where, after an uncomfortable hands and knees crawl, I could sit within a few metres of a group of guillemots. There were about twenty pairs breeding on this particular cliff edge, some of them facing out to sea as they incubated their single egg. Being so close to the birds, I had the sense of being almost part of the colony and had become familiar with all their displays and calls. On one occasion a guillemot that was incubating suddenly stood up and started to give the greeting call – even though its partner was absent. I was puzzled by this behaviour, which seemed to be occurring completely out of context. I looked out to sea and visible, as little more than a dark blob, was a guillemot flying towards the colony. As I watched, the bird on the cliff continued to call and then, to my utter amazement, with a whirr of stalling wings, the incoming bird alighted beside it. The two birds proceeded to greet each other with evident enthusiasm. I could hardly believe that the incubating bird had apparently seen – and recognised – its partner several hundred metres away out at sea.[11]

How can we establish scientifically how good avian vision is? There are two ways: by comparing the structure of their eyes with that of other vertebrates, and by devising behavioural tests to establish how well birds can see.

From the Renaissance onwards, researchers interested in human vision commonly studied the eyes of birds and other animals, and over time a picture started to emerge. Not surprisingly, it was a picture seriously biased by what was known about human vision.

Compared with mammals, birds have relatively large eyes. In simple terms, a bigger eye means better vision, and excellent vision is essential for avoiding collisions in flight, or for capturing fast-moving or camouflaged prey. Birds' eyes, however, are deceptive – they are bigger than they look. As William Harvey (famous for discovering the circulation of blood) said in the mid-1600s, birds' eyes 'outwardly appear small, because excepting the pupils they are wholly covered with skin and feathers'.[12]

As with many organs, the eyes of larger birds are generally bigger than those of smaller ones – obviously. The smallest eyes are those of hummingbirds, the largest are those of the ostrich. Those who study eyes use the distance between the centre of the cornea and lens to the retina at the back of the eye (the eye's diameter) as a measure of eye size. The ostrich's eye has a diameter of 50 mm, more than twice that of the human eye (24 mm). In fact, relative to their body size, the eyes of birds are almost twice as large as those of most mammals.[13]

Frederick II was an astute observer and in his manuscript on falconry, he commented: 'Some birds have large eyes in comparison with their bodies, some small, some of medium size.'[14] The ostrich may have the largest eye of any bird in absolute terms, but, for its body size, it is actually smaller than we would expect. The largest eyes relative to body size occur in eagles, falcons and owls. The white-tailed sea eagle has an eye of 46-mm diameter – not far off that of the ostrich (which is eighteen times heavier). At the other end of the scale, the kiwi has tiny eyes, both absolutely (an eight-millimetre diameter) and for its body size. To get some impression of just how tiny the kiwi's eyes are, the Australian brown thornbill (which weighs a minuscule six grams) has an eye diameter of six millimetres. If kiwis had eyes proportional to their body weight (which is about two or three kilograms), their eyes would be 38 mm in diameter (similar to a golf ball) – a huge difference. The kiwi's eyes have been described as 'as degenerate as it is possible for an avian eye to be'.[15]

The size of eyes is important precisely because the larger the eye, the larger the image on the retina. Imagine watching a 12-inch television compared with a 36-inch screen. Bigger eyes have more light receptors in the same way that larger TV screens have more pixels, and hence a better image.

Among diurnal birds, those that become active soon after dawn have larger eyes than those that become active later after sunrise. Shorebirds that forage at night have relatively large eyes, as do owls and other nocturnal species. The kiwi, however, is an exception among nocturnal birds, and, like those fish and amphibia that live in the perpetual darkness of caves, seems to have virtually given up vision in favour of its other senses.

The Australian wedge-tailed eagle has enormous eyes, both in absolute terms and compared with most other birds, and as a result has the greatest visual acuity of any known animal. Other birds might benefit from the eagle's acute vision, but eyes are heavy, fluid-filled structures, and the larger they are the less compatible they are with flight. Flying birds are designed so that their weight is distributed in such a way that it does not interfere too much with flight. A heavy head is incompatible with flight and therefore sets an upper limit on eye size. Flight, and the need for large eyes, may also be responsible for the loss in birds of teeth, which have been replaced by a powerful muscular stomach, the gizzard (which birds use to grind up their food), located near the centre of gravity in the abdomen.

For early researchers, vision posed many puzzles. One was why we see only a single image, even though we have two eyes. After all, with either eye we see a perfectly good image, but with both eyes open we see just one image.

René Descartes identified another puzzle, noticing that on cutting a square hole in the back of a bull's eye (that is, in the retina), and placing a piece of paper over the hole, the image projected on to the paper – through the eye – was upside down. Why, then, do we see images the right way up?

Writing about the eye in 1713, William Derham presented this puzzle thus:

> The glorious landskips [landscapes], and other objects that present themselves to the eye, are manifestly painted on the retina, and that not erect, but inverted as the laws of opticks require . . . But now the question is, how in this case the eye comes to see the objects erect?

He says that the Irish philosopher William Molyneux (1656–98) has the answer: 'The eye is only the organ or instrument, 'tis the soul that sees by means of the eye.'[16]

If we allow the 'soul' to be the brain, or acknowledge that the eye is merely an 'instrument', then Molyneux is correct. It is indeed the brain that sorts these things out, 'seeing' only a single 'erect' image. Amazingly, we train ourselves to 'reverse' the inverse image on our retinas. In a famous experiment conducted in 1961, Dr Irwin Moon wore image-inverting spectacles that effectively turned the world upside down. At first he found it horribly disorientating, but after eight days of wearing the spectacles Dr Moon had adjusted and 'saw' the world the right way up again. To prove it, he drove his motorbike and took his plane for a spin – without mishap. Moon's extreme experiment provided irrefutable evidence that we 'see' with our brain rather than with our eyes.[17]

Although we tend to think of the brain as a discrete organ – a lump of squidgy tissue – it is better to think of it as part of an elaborate network of nervous tissue that reaches out to every single part of the body. Imagine the entire nervous system: the brain, the cranial nerves emanating from it, the spinal cord, with its pairs of nerves sprouting from each side, branching and branching again, becoming finer and ever finer – dendritic is the word – with the various sense organs at their tips. Information, gathered by the sense organs, the eyes, the ears, the tongue and so on, including light, sound waves and taste, is transformed into a common currency of electrical signals that travel along the neurons to the brain, where they are decoded.

How does a duck, whose eyes are located on the sides of its head, see the world: does it see one or two images? Does a tawny owl, whose two enormous eyes face forwards like our own, see a single image as we do? Graham Martin at the University of Birmingham, in the UK, has spent many years measuring the 3-D visual fields of different bird species, and identified three broad categories of visual field.

Type 1 is what the typical bird, such as blackbirds, robins and warblers, sees: some forward view, excellent lateral vision, but (like us) no vision behind them. Surprisingly, the majority of birds in this group cannot see their own bill tip, but have just enough binocular vision to be able to feed their chicks and construct a nest.

Type 2 includes birds like ducks and the woodcock, whose eyes are high up on the sides of the head. They don't have a great forward view and most don't need to see their bill tip because they rely on other senses when feeding, but they do have panoramic vision, above and behind – helping them detect potential predators. Interestingly, the views from each eye barely overlap, so they probably see two separate images.

Type 3 birds are those such as owls with forward-facing eyes like ourselves, which have no vision behind. Because we rely so much on binocular vision for depth and distance perception, we automatically assume that all other organisms benefit in the same way. Our reliance on binocular vision may be one reason we have endowed owls with such symbolic significance, for they can look us in both eyes, with both of their eyes. But looks can be deceptive, and in fact owls' eyes are much more angled with respect to each other than they appear, and their binocular overlap much smaller than our own. It has often been thought that the forward-facing eyes of owls is an adaptation to nocturnal living, but this isn't so. Many owls are nocturnal, of course, but having a Type 3 visual field is not tightly associated with operating in the dark: oilbirds and nightjars are nocturnal, yet they have a Type 2 visual field. Martin has an interesting suggestion for why the eyes of owls face forwards. He thinks that it is associated with their need for very large eyes – associated with flying around in poor light – which, together with the need for very large external ear openings, means (as we'll see in the next chapter) that the only possible place in the skull is in a forward-facing position. 'Where else could they go?' he asks. The lack of space for both eyes and ears (and brain) in the skull is illustrated by the fact that you can see the back of an owl's eyes through its ear openings![18]

Readers of my generation educated in the UK in the 1960s will remember having the basic structure of the human eye drummed into them at school from an early age: a ball-shaped organ roughly

2.5 cm in diameter; an opening (the iris) through which light enters; a lens that projects on to the retina, a light-sensitive screen at the back of the eye. Information from the retina is transmitted via a network of nerves through the optic nerve to the visual centres of the brain. We even dissected bulls' eyes at what now seems like a very tender age: I was hooked!

When researchers first started to look into the eyes of birds and compare them with our own, they noticed a few striking differences. The first was that those of certain birds – like large owls – are more elongated than our own. The great nineteenth-century ornithologist Alfred Newton (1829–1907) described the eyeball of a bird as like 'the tube of a short and thick opera glass'.[19] The second difference is that birds possess a translucent additional eyelid, whose existence was known for centuries by everyone who kept birds. Aristotle mentions it, as does Frederick II in his falconry manual: 'for cleaning the eyeball there is provided a peculiar membrane that is quickly drawn across its anterior surface and rapidly withdrawn'.[20] The first formal description of this additional eyelid was – unexpectedly – of a cassowary, a gift to Louis XIV, that died in the Versailles menagerie in 1671.[21] John Ray and Francis Willughby in their encyclopaedia of birds of 1678 say: 'Most, if not all birds, have a membrane of nictation . . . where withal they can at their pleasure cover their eyes, though the eye-lids be open . . . and serves to wipe, cleanse, and perchance moisten . . .' The term nictitating membrane comes from the Latin *nictare*, to blink. Our own nictitating membrane is a mere remnant – the tiny pink nub in the inner corner of our eye.[22]

A bird's nictitating membrane lies under its other eyelid and is most easily seen in photographs. If you have ever taken close-up pictures of birds at the zoo I bet you have images in which the bird's eye seems to be milky or obscured in some way, even though it looked all right as the picture was being taken. Usually the milkiness is caused by the nictitating membrane moving rapidly across the eye, either horizontally or obliquely, in a movement almost too quick to be seen, but readily captured by the camera. As Frederick II recognised, the nictitating membrane's function is to clean the eye, but it also protects it. Each time a pigeon puts its head down to peck at something on the ground, the nictitating membrane moves across each eye to protect it from spiky leaves and grasses. In raptors the membrane covers the eye immediately before the bird slams into its prey, and in exactly the same way the membrane covers the eye just before a plunging gannet hits the water.

The third difference between our eyes and those of birds is a structure called the pecten. So-named because of its resemblance to a comb (in Latin, *pecten*), the pecten seems to have been discovered in 1676 by Claude Perrault (1613–88), one of the great anatomists of the French Academy.[23] The pecten is a very dark structure with a pleated appearance, with the number of pleats varying between three and thirty in different species. At one point ornithologists hoped – as they had with so many other anatomical traits – that the pecten might provide vital information about the relationships between species. It didn't. The pecten is, however, largest and most complex in those birds with the most acute vision, like raptors. Indeed, it was initially thought that the kiwi lacked a pecten altogether, but in the early 1900s Casey Wood found that it possessed a small and very simple one.[24]

At first sight the pecten looks as though it would impede rather than improve vision, sticking out like a large finger inside the rear chamber of the eyeball. Yet, on closer inspection, anatomists – including Casey Wood – realised that it is cunningly positioned so that its shadow falls on the optic nerve – or blind spot of the retina – and therefore does not interfere with vision. What is the pecten for, and why don't we have one? The pecten in birds seems to be to provide the posterior chamber of the eye with oxygen and other nutrients. In contrast to humans and other mammals, there are no blood vessels in the avian retina, and the pecten, which is a mass of blood vessels, is little more than a clever oxygenation device – the pleating maximising its surface area and thereby facilitating the exchange of gases (oxygen in and carbon dioxide out) within the eye – effectively allowing it to breathe.

The human fovea – the crucial spot on the back of the eye where the image is sharpest – was discovered in 1791. Over the following years foveas were found in a wide range of other animals, but it wasn't until 1872 that they were discovered in birds.[25] Not long afterwards it was noticed that, while the majority of birds have a single circular fovea – as we do – some, like hummingbirds, kingfishers and swallows, as well as raptors and shrikes, have two. Remarkably, a few species, including the domestic fowl, have no fovea at all. Others have a linear fovea, and yet others have some combination of the two. Many seabirds, including the Manx shearwater, have a linear, horizontal fovea whose function may be to detect the horizon.

In birds like falcons, shrikes and kingfishers, the two foveas are referred respectively to as the shallow and deep fovea.[26] The shallow fovea is like that in birds with only a single fovea and provides monocular, and largely close-up, vision. The deep fovea, however, which is directed at about 45° to the side of the head, comprises a spherical depression in the retina that acts like a convex lens in a telephoto lens, effectively increasing the length of the eye and magnifying the image to provide very high resolution.[27] The position of the deep fovea in the eye also means that raptors have a degree of binocular vision, thought to be essential for judging the distance of fast-moving prey.[28] If you have observed captive birds of prey, you will see that they often move their head from side to side or up and down as they watch you approach. They do this because they are alternating your image on their two foveas, the shallow one for close up, the deep one for distance. Compared with our eyes, those of birds are relatively immobile in their sockets (space and weight are limited, and the reduction of muscles needed to move the eyes constitutes an important saving), so raptors and owls in particular have to move their head when they are scrutinising something.

The size and basic design of birds' eyes can tell us only so much, but the microscopic structure of the retina is more revealing. The wonderful visual acuity of raptors is largely the result of a high density of light-sensitive cells in the retina. The light-sensitive cells, or photoreceptors, come in two main types: rods and cones. Rods can be thought of as working like old-fashioned high-speed black and white film – capable of detecting low levels of light. Cones, on the other hand, are like low-speed (ISO) colour film (or a low ISO setting on a digital camera) – high-definition and performing best in bright light.

Our own single fovea is defined by a slight depression in the retina where the density of cone photoreceptors is very high, and where each photoreceptor has its own nerve cell sending information to the brain. Elsewhere in the eye each photoreceptor cell (i.e. both rods and cones) shares nerve cells, rather like lots of people having their computers connected to the internet via a single telephone line – frustratingly slow. The one-to-one relationship between photoreceptor and nerve cells in the fovea means that each cone sends an independent message to the brain, providing a signal whose origin is more accurately located, and explains why the fovea is the region of maximum resolution and colour imaging.

What a bird sees is dictated by the gross structure and size of the eye, the density and distribution of photoreceptors in the retina, and the way the brain processes the information transmitted through the optic nerve. Although all three aspects are correlated, any one of them on its own provides only a poor indication of a bird's visual sensitivity, or how much detail a bird can see.

The raptor eye has excellent visual *acuity* – the ability to see fine detail. The owl eye, on the other hand, has excellent *sensitivity* – the ability to see at low light levels. No eye can do both, for the same reason that a camera cannot simultaneously have a wide aperture and a great depth of field. It is simply a law of physics. As vision biologists Graham Martin and Dan Orsorio say: 'There's always a trade-off between these two fundamental visual capacities [sensitivity and acuity]: if there are few quanta in the image [little visual information because the light is poor] the resolution cannot be high, and if the eye is designed to achieve high spatial resolution, it cannot do it at low light levels.'[29] Visual acuity depends on the basic design of the eye, including its size (because this dictates the size of image projected on to the retina), and the design of the retina itself. The situation is analogous to a camera: the quality of lens determines the quality of the image, and the speed (grain) of the film (or the ISO setting on a digital camera) determines the accuracy with which the image is reproduced. Raptor retinas have a preponderance of cones, especially in each fovea, where the density is about one million cones per square millimetre (compared with some 200,000 in humans). As a result, a raptor's visual acuity is slightly more than twice as good as our own.

ϵₜᵗϵₗ

Birds are among the most colourful of animals, which is, of course, one reason we find them so appealing. One of the most brilliantly coloured of South American birds (and there are many) is the Andean cock-of-the-rock. The male has the most intensely red body, a jet-black tail and outermost wing feathers, and unexpectedly silvery-white innermost wing feathers. So-named because it nests among rocks on cliff ledges, and because of its cocky Mohican-like crest, this pigeon-sized bird is a major draw to birdwatchers visiting Ecuador. The males display in groups, referred to as 'leks', deep in the rainforest, and it was with a group of some fifteen or so other birders that we made our way down a steep, slippery track towards a display area. Long before we saw them, the birds announced their presence with distinctive screeches, which the local Quechua people render as *youii*.

From the viewing platform on the valley side, the birds were surprisingly difficult to see. The vegetation was dense, and although the males were actively chasing each other from tree to tree, they came into view only occasionally and rarely remained long enough in one place to register a satisfying image on my retina. I kept willing them to perch in the sun so that I could see them properly. Eventually when one did, it was stunning and put me in mind of a fleck of glowing volcanic lava amidst a mass of green foliage.

The most memorable thing about my brief encounter with the cock-of-the-rock was that, despite the birds' brilliant colour, as soon as they moved out of the sun they became almost invisible. It was like watching an actor step from out of a spotlight into the darkness, and disappear. This effect is no accident. Males choose sunny display sites to maximise the wonderful effect of their plumage. Evolution has designed these birds such that when illuminated by the sun they appear utterly brilliant, but in the shade, with the light filtered through green forest vegetation, their plumage has an almost drab quality, rendering the bird surprisingly well camouflaged.

As I watched the males flitting from perch to perch in the dense foliage, I wondered how the ornithological pioneers ever worked out what was going on at the cock-of-the-rock lek: I didn't see a female, and consequently never saw the males in full courtship mode. Local people had obviously known about the birds and their leks for millennia, and used the males' scarlet feathers in their headdresses.

The first description of a cock-of-the-rock lek came from Robert Schomburgk, a geographer charged by Queen Victoria with the daunting task of mapping British Guiana (now Guyana). On 8 February 1839, during a tough day of climbing as he crossed the mountains between the Orinoco and the Amazon, Schomburgk and his colleagues watched a group of ten males and two females: 'The space was four or five feet in diameter, and appeared to have been cleared of every blade of grass and smoothed as though by human hands. A male was capering to the apparent delight of the others.' In 1841 Schomburgk's brother, Richard, a botanist and ornithologist, went back and confirmed Robert's extraordinary observations. On hearing the cries of the cock-of-the-rock, 'My companions immediately sneaked with their weapons in its direction, when soon after one of them returned and told me to follow him carefully and lightly. We might have crept some thousand paces through the bush on our hands and knees when . . . on crouching down quietly besides the Indians, I witnessed the most interesting sight.' A lek in all its glory, with birds 'uttering the most peculiar notes . . . one of the males was cutting capers [dancing] on the smooth boulder; in proud consciousness of self it cocked and dropped its widespread tail and flapped its likewise expanded wings . . . until it seemed exhausted, when it flew back on the bush'.[30]

Like several other lekking bird species, male cock-of-the-rock choose their display sites with great care. The satin bowerbird of Australia selects sun spots, but some birds of paradise in New Guinea and manakins in South America actually create their own sun spot on the forest floor by pruning adjacent trees. It was once thought that this 'gardening' was to minimise the risk of predation, but as our understanding of avian vision improved, it became clear that the birds were manipulating the background colour to maximise the visual contrast of their plumage and the overall effectiveness of their sexual displays.

I was thrilled by the sight of male cock-of-the-rock and their brilliant colour in the sun, but I wondered whether a female would see them as I did. In fact, as we'll see, females see them even more brilliantly.

As Darwin recognised, the bright colours of male birds, like those of the cock-of-the-rock, were unlikely to have evolved because they enhanced survival. Instead, such traits must have evolved because they increased reproductive success. Darwin imagined this happening in one of two ways: either males competing among themselves for females, or females preferentially mating with the most attractive males. It was an ingenious idea and neatly accounts for what are often dramatic differences in appearance and behaviour of the two sexes. Darwin called it sexual selection, to distinguish it from natural selection, recognising that even if bright plumage or loud songs rendered males more vulnerable to predators, if they were attractive enough to females and left enough descendants, they would still be favoured by selection. There were problems, though, especially with the second process, of female choice. Darwin's contemporaries simply couldn't imagine that females (human or non-human) were smart enough to make such informed choices. But by imagining that such choice required consciousness, they missed the point. A more serious problem was one raised by Alfred Russel Wallace, who pointed out to Darwin that he had not said *how* females benefitted from mating with particularly attractive males. Indeed, Darwin did not know.

These two objections effectively killed off the study of sexual selection, and in the several decades following Darwin's death few researchers bothered to pursue it. Remarkably, it was not until a major shift in evolutionary thinking in the 1970s that female choice became scientifically respectable again. The turning point was the recognition that selection operated on individuals rather than groups or entire species, and that as a consequence females could benefit in several different ways by choosing to mate with particular males. In the case of species like the cock-of-the-rock, where males make no material contribution to offspring other than through their sperm, the most likely benefits females obtain from choosing particular males is the acquisition of better genes for their offspring.[31]

To understand *how* females choose between different males, researchers in the last decade or so have started to consider the avian sensory system. In the case of the cock-of-the-rock one would need to see the world – or to see males at least – through a female's eyes. While we cannot do this literally, we now know enough about how birds' eyes work to be able to make a well-informed guess, simply (well, not so simply, actually) by looking at the microscopic struc-ture of their eyes. The reason why this has been such a major step forward is that we now know that colour is a property of both an object, such as a bird or a feather, and of the perceiver's nervous system that analyses its image thrown upon the retina. Beauty is, indeed, partly in the eye of the beholder: in fact, in the *brain* of the beholder, for that's where images are processed. Without knowing about the nervous system we cannot really grasp how birds might 'see' each other, or, indeed, how they see the environment in which they live. It has taken a surprisingly long time to realise this, and as Innes Cuthill at the University of Bristol, in the UK, has said, while we readily accept that a dog has a much better sense of smell than we do, we have been incredibly reluctant to accept that birds, or any animals, *see* the world differently from ourselves.

Let's consider the photoreceptors (cones) in the retina responsi-ble for colour. Humans have three types, defined by the colour of the light they absorb: red, green and blue. These are directly equiva-lent to the three colour 'channels' on a television or video camera, which in combination produce what we consider to be the full spec-trum of colour. Compared with most mammals, humans and primates have relatively good colour vision, because most others

– including dogs – have only two cone types, which must be like having only two (instead of three) colour channels on a television. However good we (arrogantly) think our colour vision is, compared with that of birds it is rather poor because they have four single-cone types: red, green, blue and ultraviolet (UV). Not only do birds have more types of cone than ourselves, they have more of them. What's more, birds' cone cells contain a coloured oil droplet, which may allow them to distinguish even more colours.

The UV cone type in birds was discovered only in the 1970s. Prior to that, UV vision had been known in insects since the 1880s when Darwin's neighbour John Lubbock noticed it in ants. Just a few decades later biologists discovered that honeybees use UV vision to discriminate between flowers. In the mid-twentieth century UV vision was assumed to be limited to insects, providing them with a private communication channel invisible to predators like birds.

This was wrong, and a study of pigeons in the 1970s showed that they were sensitive to UV light. It is now known that many birds, probably most,[32] have some degree of UV vision that they use to find both food and partners. The berries that some birds feed on have a UV bloom; and European kestrels can track their vole prey from the UV reflecting off the voles' urine trails. The plumage (or parts of it) in hummingbirds, European starlings, American gold-finches and blue grosbeaks reflect UV light and often more markedly in males than females. In some species, like the blue grosbeak, the degree of UV reflectance also reflects male quality, so it is little wonder that females use this aspect of plumage to discriminate between potential partners.[33]

Most owls are nocturnal. Good night vision is therefore essential, mainly for negotiating obstacles rather than locating prey, since owls hunt mainly using their ears. The key issue for nocturnal owls is the sensitivity of their eyes. To establish the minimum amount of light they can detect, Graham Martin conducted some behavioural tests with tame tawny owls – one of only a handful of species for which such information currently exists. Over several months, the owls were trained to peck at a bar positioned in front of two screens through which lights of different intensity were projected. The birds were rewarded with a bit of food if they detected the light. Martin used exactly the same procedure (but without the food reward) with human subjects so that he could make a direct comparison. As we might expect, the owls were more sensitive than the human subjects and on average could detect much lower light levels than most humans, although a few human subjects were more sensitive than the owls.[34]

The eyes of a tawny owl are enormous compared with those of most other birds, and in terms of their focal length they are remarkably similar to the human eye (they both have a diameter of about 17 mm). However, because the owl's pupil is larger (13 mm in diameter) than a human's (8 mm), it lets in more light, and the image on the owl's retina is more than twice as bright as it is in humans – accounting for the difference in visual sensitivity. Tawny owls are woodland birds and Martin checked whether there were ever conditions when there would be insufficient light for them to operate efficiently. He found, not surprisingly, that under most circumstances there was sufficient light and only when the owls were under a dense tree canopy on a moonless night would they struggle to see clearly.

Comparisons with a bird that is strictly diurnal, the pigeon, for example, shows that the tawny owl's sensitivity to light is about a hundred times that of the pigeon. That is, owls see much better in poor light than pigeons, and this explains how owls function so well at night. In full daylight, both the pigeon and tawny owl have similar levels of visual acuity, confirming that, contrary to what some people believe, tawny owls are at no disadvantage in daylight.

Because the owl's eyes are designed for maximum sensitivity rather than resolution they can see quite well at low light levels, but not very crisply. By comparison, the spatial resolution – the ability to discriminate fine detail – of diurnal raptors such as the American kestrel and the Australian brown falcon, is five times greater than that of the tawny owl.[35]

The fact that birds use their right and left eyes for different tasks is one of the most extraordinary ornithological discoveries of recent times. As in humans, a bird's brain is divided into two hemispheres, right and left. Because of the way the nerves are arranged, the left half of the brain processes information from the right side of the body, and vice versa. That different sides of the brain deal with different types of information was first recognised in the 1860s by the French physician Pierre Broca, after examining a man with a speech defect and whose subsequent autopsy revealed that the left hemisphere of his brain was severely damaged – as a result of syphilis. A gradual accumulation of similar cases confirmed that the left and right hemispheres of the brain do indeed process different kinds of information. The effect is called 'lateralisation' – meaning 'sidedness' – and for a century or so was thought to be unique to humans. But in the early 1970s, during a study of how canaries acquire their song, it was discovered that birds, too, have a 'lateralised brain'. In canaries and other birds, their song emanates from the syrinx, a structure similar to our voice box. Fernando Nottebohm found that the nerves on the left side of the canary's syrinx (and hence the right side of the brain) had no role in song production, whereas those of the right did – providing an important clue that song acquisition in birds, like human language, was more dependent on one side of the brain than the other. Subsequent research confirmed that this was exactly the case.[36]

More than that, birds have continued to play a central role in understanding brain lateralisation and it is now recognised that sidedness in brain function enhances the processing of information, effectively allowing individuals to use several sources of information simultaneously.

Sidedness can be apparent into two different ways. First, in terms of the *individual*: humans, parrots and some other animals can exhibit sidedness, being either left- or right-handed or -footed (parrots). Second, entire *species* can exhibit sidedness, as in domestic fowl, which, as we'll see, typically use their left eye to scan for aerial predators.[37]

Humans are, of course, typically right- or left-handed; we also tend to have a dominant eye – in about 75 per cent of people it is the right eye – although we are not usually aware of using our eyes differentially. Yet in those birds whose eyes are placed 'laterally', that is, on the side of the head, the two eyes are used for different tasks. Day-old chicks of the domestic fowl, for example, tend to use their right eye for close-up activities like feeding and the left eye for more distant activities such as scanning for predators. What's more, an ingenious behavioural test, in which one eye is temporarily covered with a patch, reveals that birds perform certain tasks much better with one eye than the other, including tits and European jays remembering where they've hidden food.[38]

We even know how this differential use of each eye arises in birds. The leading researcher on lateralisation in birds, Australian Lesley Rogers, had often wondered how the phenomenon arose. Lesley told me this:

> All of my colleagues assumed it was determined genetically but I was not so sure. Then, one day [in 1980] I was looking at photos of the chick embryo and noticed that, during the final days of incubation, the embryo turns its head to its left

side so that it occludes [covers] its left eye but not its right eye. That gave me the idea that light reaching the right eye via the shell and membranes might establish visual lateralization. Therefore, I compared eggs incubated in darkness with those exposed to light for the last few days of incubation and showed that my idea was correct. Later I showed that you can even reverse the direction of lateralization by removing the late-stage embryo's head from the egg and occluding the right eye while exposing the left to light.[39]

It is remarkable that the difference in the amount of light each eye receives during normal embryonic development (left: rather little; right: much more) determines the subsequent role of each eye. The chicks hatching from eggs that have been experimentally allowed to develop in complete darkness (so that there is no right-left bias in the amount of light each eye receives) show no such difference in eye use once they hatch. What's more, those chicks were less competent at performing two tasks simultaneously (detecting predators and finding food) than chicks hatching from eggs incubated normally.[40]

This remarkable discovery has enormous and as yet unexplored implications. Imagine a hole-nesting species which sometimes nests in deep, totally dark cavities, but occasionally nests in a shallow, light-filled cavity. In the first case there would be no opportunity for lateralisation, whereas in the second there would, and as a consequence the offspring would be of better 'quality' – because they would be more competent. If this is true, then differences in the environment in which they were raised could explain a lot about individual differences in behaviour and personality in birds. We might almost expect individuals to advertise – through display – how lateralised they are, since highly lateralised, more competent individuals will inevitably make better partners. A wonderful project for a budding ornithologist!

This bias in the role of each eye is difficult for us to imagine, but it may occur in all birds, albeit in different ways. Domestic fowl chicks, for example, use their left eye to approach their parent. Male black-winged stilts are more likely to direct courtship displays towards females seen with their left eye than with their right. The wrybill, a New Zealand plover, is unique among birds in having its bill curved laterally to the right, which it uses to flip over stones as it searches for invertebrates – either because the right eye is better for close-range foraging or because the left is better for spotting potential predators, or both. When peregrine falcons are hunting they home in on their prey in a wide arc, rather than in a straight line, and mainly use their right eye.[41] New Caledonian crows, famous for their construction of tools – making hooks from palm leaves – show a strong individual bias towards making tools from either the right or left side of leaves. Similarly, when actually using these tools to hook prey out of crevices they show an individual preference for their left or right side, but no bias exists towards left or right in the population as a whole.[42]

Given how widespread sidedness is, it is natural to assume that it has a function. And indeed it has. Intriguingly, the more biased the sidedness is (at both individual and species level), the more proficient those individuals are at particular tasks. It has been known for centuries that parrots consistently prefer to use one foot to grasp food or other objects. The more biased parrots are towards using one particular foot (and it doesn't matter whether this is the left or the right) the better they are at solving tricky problems – like how to pull up a food reward dangling from the end of a string. The same thing is true of fowl chicks – those with strong sidedness are much better at foraging (discriminating between food grains and gravel) *and* keeping an eye open for predators in the sky.[43]

We'll end this chapter by looking at how and why some birds are apparently able to sleep while still looking at the world through one eye. This ability was recognised as long ago as the fourteenth century when Geoffrey Chaucer in *The Canterbury Tales* (1386) wrote: '. . . smale fowles . . . slepen [sleep] at the night with open ye . . .' Sleeping with one eye open is something we now know birds share with marine mammals (which need to return to the surface to breathe), but certainly not with us.[44] It is not even true of all birds, and so far it is known that songbirds, ducks, falcons and gulls can sleep with one eye open, but a complete survey has yet to be undertaken. One-eyed-sleep is easiest to see in ducks roosting during the day beside urban ponds: with its head turned back towards the wing (often incorrectly described as 'with its head under its wing'), the bird has one eye facing inwards towards its back and concealed, the other eye looking outwards and opening from time to time.

As you will probably have guessed, a bird sleeping with its right eye open is resting the right hemisphere of its brain (since information from the right eye is processed in the left hemisphere and vice versa), and there are two circumstances in which the ability to sleep with an eye open is incredibly useful. The first is when there is a predator about. Ducks, chickens and gulls often sleep on the ground and are vulnerable to predators like foxes, so it pays to keep one eye open. A study of mallard ducks showed that individuals sleeping in the centre of a group (where it is relatively safe) spent much less time with an eye open than those on the edge (where they are more vulnerable to predators), and that ducks on the edge of the group were more likely to open the eye facing outwards from the group in the direction from which a predator might approach.[45]

The second circumstance in which it is extremely useful for birds to keep an eye open is when they sleep on the wing – that is, while flying. The idea that birds might sleep and fly simultaneously once seemed ludicrous, but was considered more than just a possibility by the ornithologist David Lack when he was studying European swifts. He and others noticed swifts ascending into the sky at dusk and not returning until the following morning, and inferred that they must sleep on the wing. More convincingly, a French airman on a special nocturnal operation during the First World War reported that, as he glided down across enemy lines with his engine off, at an altitude of around 10,000 ft: 'We suddenly found ourselves among a strange flight of birds which seemed to be motionless . . . they were widely scattered and only a few yards below the aircraft showing up against a white sea of cloud underneath.' Remarkably, two birds were caught and identified as swifts. Of course, neither Lack nor the French airman noticed whether their sleeping swifts had one eye open, but it is a possibility. Glaucous-winged gulls in North America, however, have been seen flying to their roosts with only one eye open, suggesting that they are already sleeping before even reaching the roost.[46]

Rather than end this chapter on a sleepy note, I want to finish on something a bit more dynamic – the extraordinarily rapid flight of certain birds. Think of a descending swift; or the way a hummingbird zips from one bloom to another; or the way a sparrowhawk or sharp-shinned hawk hurtles among the branches after its prey. Such high-speed movements must require high-speed brain function, and I've often wondered how birds do it. Perhaps we shouldn't be too surprised that birds have this ability since insects, whose brains are much smaller and whose vision is much less sharp, manage extremely well.

The closest we can come to imagining what it is like to process information as rapidly as a hummingbird or a hawk is the sensation of time slowing down that occurs during a near-death experience. Over the years I have had a few near-death experiences while doing fieldwork, and I imagine many readers will have had, like me, the same sensation during traffic accidents. As you slam on the brakes and slide inexorably towards another vehicle or a tree, it is as if your brain is taking in every detail and each second is drawn out until it feels ten times longer than it really is.

The bizarre thing is that, while this provides a convenient way for us to imagine what it is like to be a fast-moving bird, psychologists now realise that the sensation of time slowing down in near-death situations is an illusion. It is a quirk of our memory: scary events are remembered in great detail, so we perceive a slowing down of time only *after* the event. The hummingbird or accipiter, of course, experience events in real time.[47]

2

Hearing

The great grey owl – with its enormous facial disc for sound collection. Thumbnails are of a saw-whet owl (from left to right): left ear opening; the skull with its aymmetric ear openings (more asymmetric than any other owl), and the location of the ears (arrows)..

It cannot be doubted that the faculty of hearing is highly developed in birds, not only the mere perception of sound, but also the power of distinguishing or understanding pitch, notes and melodies, or music.
From Alfred Newton, 1896, *A Dictionary of Birds*, A. & C. Black

This is a strange place: it is dark, wet and by British standards curiously remote. The horizon of the night sky is stained orange by the urban glow of Peterborough and Wisbech, while somewhat closer the floodlit chimneys of a brickworks belch fiery columns of smoke into the clouds. On the flat, featureless landscape I see the lights of an occasional car trundling along quiet country roads. The most bizarre aspect of this place, though, is the repetitive, toneless *crex crex* of corncrakes from the black meadows. One bird is fairly close, another more distant, but it is difficult to tell for the call has a curious ventriloquial quality: sometimes loud, sometime quiet, depending on which direction the birds are facing.

Hoping simultaneously to deter male corncrakes and attract a female with his mechanical rasping call, this bird – not much larger than a thrush – could live out an entire breeding season without ever being seen by a human. This is a bird whose presence is betrayed only by its voice.

Looking out across the Nene Washes I see there are houses whose bedroom lights are on and windows open. I imagine people lying in bed and hearing the corncrakes: do they recognise this reassuring ornithological renaissance for what it is?

Corncrakes once thrived here before the Washes were drained by ingenious Dutch engineers drafted in for the job. Back then this area was one huge wetland heaving with insects, birds and other wildlife. Even now, repaired and restructured by the Royal Society for the Protection of Birds (RSPB) and others, the Washes hold some special birds including spotted crakes, cranes, black-tailed godwits, ruffs and snipe.

As we plough our way through waist-high grass wet from an earlier downpour, the air becomes heavy with the scent of water mint. A corncrake calls nearby, or so it seems. 'Here,' says Rhys: 'This is where we will put the net.' We erect the 60-ft mist net with hushed voices and muted head torches. Like some weird clockwork toy the corncrake continues to call, apparently oblivious of our efforts. Rhys, with a tape recorder crudely packaged in polythene bags to protect it from the wet, positions himself behind the net and in line with the bird, and I creep with my tape recorder into position between the bird and the net in case it overshoots as it tries to confront the auditory intruder and has to be called back.

Rhys is the RSPB's corncrake champion, and for several years he has been overseeing the species' reintroduction to this part of England. We are old friends, having first met at a student bird conference back in 1971. Rhys's recorder blasts out a crude, almost deafening echoey corncrake: recorded elsewhere during the day, there's a skylark trilling between the booming *crexes*.

It is a continuous, relentless loop, much like the programme in the bird's brain seems to be. I cannot imagine what's running through the real bird's head, but suddenly it stops calling; there's a barely audible flutter overhead as it launches itself at the apparent intruder, and it is in the net. 'Right!' shouts Rhys and we spring into action to retrieve the bird. Reaching inside the folds of the net, I can see that the bird is already ringed. It is in fact one of several captive-reared corncrakes released earlier in the year. In the hand this is a beautiful russet and grey bird, whose laterally compressed body and wedge-shaped head are beautifully designed for pushing through the grass. Quickly checked and weighed, the bird is released and we walk back to the car.

Driving along a pockmarked road, dodging the enormous puddles, we stop and through the open windows listen again. 'There's one,' Rhys says and we collect the net and walk out across the sodden fields towards the sound. The protocol is as before, with me between the bird and the net. On goes the tape, blasting its challenge across the flat, wet landscape. The territory owner continues to rasp. On and on the tape continues; on and on goes the bird: stalemate, I think. It is uncomfortable lying in the grass, the tips of the blades are tickling my nose, neck and face, but I dare not move. The bird stops. Has it given up, defeated by its much louder rival?

All at once I am aware of a sound in the grass, almost like the footfall of distant cattle. Then it stops. An illusion? I'm not sure. The rustling starts again and I realise the corncrake is walking towards me. Almost unbelievably just a few centimetres from my head, but totally invisible, he starts to call again. At point-blank range the full power of his *crex* is even louder than the tape. He is moving again, and very close. Against the glow in the night sky I can see the seed heads of the grass wobbling. All at once he is walking past my face: a flurry of wings and he's in the air – and in the net.

'Right!' shouts Rhys, jolting me out of my reverie, and it is straight down to ringing. This bird is not ringed and is therefore entirely wild, evidence that the captive-reared birds are doing their job and successfully drawing in migrant concrakes. In the hand the bird is accommodating and patient. A few minutes of processing, the only real insult the glare of our head torches, and he is gently released back at exactly the spot where we first heard him. And a minute later, he's found his tape loop and is off again in his relentless quest to attract a female.

I discover later that at close range the corncrake's call registers about 100 decibels (dB). Putting that in context, normal conversational speech at the same distance is about 70 dB; a personal stereo at maximum volume is around 105 dB and an ambulance siren about 150 dB. Fifteen minutes of corncrake calls at this close range and I'd start to damage my ears.

Why, then, doesn't it damage the corncrake's ears? After all, the corncrake is even closer to its own call than we could ever be. The answer is that birds possess a reflex that reduces the sound of their own voice. This auditory reflex may be extreme in the case of the capercaillie, a turkey-sized game bird, in which the male performs a particularly noisy courtship display. The nineteenth-century ornithologist Alfred Newton wrote this about it: 'It is well known that the cock for several seconds towards the end of his rutting ecstasy is completely deaf to any external sounds.'[1] According to the German ornithologists who investigated the underlying mechanism in the 1880s, the male capercaillie's temporary deafness is the result of the external ear being blocked by a flap of skin while he calls and for a few seconds afterwards. Subsequent studies of a range of bird species suggest that simply opening the mouth wide to call results in a change in tension on the eardrum, reducing the ability to hear.[2]

Despite its mechanical, toneless quality, the corncrake's call has the same function as the song of a passerine bird: a long-distance signal that says 'keep out' to other males, and 'come in' to females. Long-distance indeed, for the corncrake's rasping call can be heard over one mile away. While this is fairly remarkable it isn't the most extreme. The record for sound transmission goes to two birds whose deep, booming calls are sometimes audible to humans as far as two or three miles away.

The first of these is the European bittern, nicely described by Leonard Baldner, a fisherman-naturalist living on the Rhine in the mid-1600s. Baldner noted that the bittern's boom was uttered with the head held high and the bill closed and that the bird has 'guts with a long stomach five ells long' (an ell is an old unit of measurement), referring to the bittern's enlarged oesophagus which is employed in sound production.[3]

The second is the kakapo, New Zealand's flightless giant parrot, whose booming was familiar to the Maori at the time of the first European settlers: 'At night . . . the birds come forth and collect at their . . . common meeting place or playground . . . having collected, every bird . . . goes through a strange performance by beating its wings on the ground and uttering its weird cry, at the same time forming a hole in the ground with its beak.'[4] Writing in 1903 Richard Henry said: 'I think it likely that the males take up their places in these "bowers" [the bowls], distend their air sacs, and start their enchanting love songs; and that the females . . . love the music . . . and come up to see the show.'[5] By watching through a night-vision scope New Zealand's kakapo hero, Don Merton (1939–2011), confirmed that males assume an almost spherical shape during their booming.[6] Unlike the corncrake, which, like most other birds, relies mainly on his syrinx (or voice box) to make his presence known, both the bittern and probably the kakapo, too, use their oesophagus, gulping down air and then releasing it in a booming belch.

Predominantly nocturnal, the corncrake, bittern and kakapo all live secretive lives amidst dense vegetation, relying on their loud calls to announce their presence and on their hearing to detect the presence of others.

Long-distance communication is not, of course, restricted to nocturnal birds; most small birds sing to advertise themselves to potential territory intruders and to potential partners, and benefit from having their song heard from as far away as possible. One of the loudest of all songbirds is the nightingale and I once spent an almost sleepless night in a small bed and breakfast on a wooded hillside in Italy being 'serenaded' ('blasted' would be a better term) by a male just one metre from my bedroom window. It was so loud I could feel his song resonating in my chest! Laboratory studies show that nightingales sing at around 90 dB.[7]

If we want to know what a human can hear, we simply ask. To establish what birds can hear we have to ask in a rather different way. This is most often done by looking at their behavioural responses to sounds, typically using captive birds such as the zebra finch, canary and budgerigar as 'models' for other species. Studies of this kind involve training birds to perform a simple task like pecking a key in response to hearing a particular sound to get a food reward. If they (consistently) perform the task it is assumed they can hear the sound, or distinguish between different sounds (and vice versa).

Spelled out like this, the study of avian hearing seems straightforward, but our understanding of hearing in birds still falls far short of what we know about vision. This is partly because birds have no external ears and because (as in most vertebrate animals) the most important part of the ear is deeply embedded in the bones of the skull. But perhaps most significant of all, there has simply been much less interest in hearing than in vision. At the time that John Ray and Francis Willughby were writing their ground-breaking *Ornithology* in the 1670s, almost nothing was known about the structure of ears in birds. Even the great anatomists of the seventeenth, eighteenth and nineteenth centuries found dissection of the inner ear a major challenge.

The first serious investigations of the human ear were undertaken by Italian anatomists in the 1500s and 1600s. Gabriel Fallopius (1523–62) – after whom the Fallopian tube in the reproductive system of female mammals is named – discovered the semi-circular canals in the inner ear in 1561. Batholomaeus Eustachius (1524–74) – after whom the Eustachian tube is named – discovered the middle ear in 1563 (the ancient Greeks already knew of the cochlea). Giulio Casserius (1552?–1616) discovered the semi-circular canals in the inner ear in pike in 1660, and found that birds (i.e. the goose) have only a single bone (rather than three) in the middle ear. The French anatomist Claude Perrault was the first to describe the inner ear of any bird. His discovery was the result of dissecting a currasow, a turkey-like bird from tropical South America, which had died in the Zoological Gardens in Paris.[8]

That was the descriptive phase. Discovering how the ear actually worked would take rather longer. Even by the 1940s when Jerry Pumphrey (1906–67), a lecturer at Cambridge University, wrote a short but seminal overview of the senses of birds in 1948, he summed up by saying: 'It will have been noted that there is a sufficient body of knowledge of the avian eye to permit of intelligent speculation about its performance and the part it plays in avian behaviour. This is far less true of the ear ... [and avian hearing offers] a most promising and unjustly neglected field for experiment and observation.'[9]

Since the 1940s there has been increasing interest in what birds can hear, driven largely by spectacular advances in the study of birdsong, that has served as a general model for learning and for understanding human speech acquisition. It was once thought that children were able to learn any language they were exposed to because they started life as a blank slate. The study of birdsong dispelled this idea by demonstrating that, although young birds are capable of learning almost any song they hear, they actually possess a genetic template that dictates both what they learn and how they sing. The study of the way birds acquire their song has provided the most compelling evidence that there is no nature-nurture divide: genes and learning are intimately interconnected in both birds and babies. It was through the study of the neurobiology of birdsong that we began to realise the huge potential for the human brain to reorganise itself and form new connections in response to particular inputs.[10]

In both birds and mammals, the latter including ourselves, the ear consists of three regions: outer, middle and inner. The outer ear comprises the auditory canal (and in most mammals an external ear). The middle ear consists of the eardrum and either one or three middle ear bones. The inner ear comprises the fluid-filled cochlea. Sound (technically, acoustic pressure) is transmitted from the environment through the outer ear, down the auditory canal and on to the eardrum, then via the tiny ear bones to the inner ear, causing the fluid inside it to vibrate. The vibrations cause microscopic hair cells in the cochlea to send a signal to the auditory nerve and then on to the brain which decodes the message and interprets it as 'sound'.

There are four main differences between human ears and those of birds. *First* and most obvious is the absence in birds of an external ear, or 'pinna' – the skin-covered bit of cartilage we call our ear.[11] It isn't always obvious where a bird's ears are because, in all but a few species, they are covered with feathers known as the ear coverts. The ear opening lies behind and slightly below the eye, in roughly the same position as our own: it is obvious if you look at the sparsely feathered head of a kiwi or an ostrich, or the naked head of New World vultures, like the condor, or the aptly named bare-necked fruit crow.[12]

In birds with feathered heads the ear coverts differ from adjacent feathers by being rather shiny, a feature that may ensure a smooth flow of air over the ears while the bird is in flight, or that may facilitate hearing by filtering out the sound of the wind passing over the ears.[13] In seabirds the feathers covering the auditory canal prevent water getting into the ear while they are diving, a potentially serious problem for species like the king penguin which dives to several hundred metres, where the pressure is considerable. In fact, the ears of king penguins exhibit a number of anatomical and physiological adaptations to protect them from the problems associated with deep-sea diving.[14] The kiwi would clearly benefit from some additional protection of its auditory canal, for several of those I handled in New Zealand had ticks lodged inside their ear openings! I later wondered whether these ticks might be an unpleasant by-product of New Zealand's relatively recent invasion by man's domestic animals and their parasites, but it appears that the kiwi ticks I saw are native to New Zealand and an inconvenience that kiwis have been coping with for a long time.[15]

In 1713 William Derham, a colleague of John Ray, noted that the 'outer shell or pinna is missing in birds, because it would impede their passage in air'. For Derham the perfect match between an organism's design (in this case the absence of a pinna) and its life-style (flight) was evidence of God's wisdom. In today's terminology, we would simply say that this was an adaptation for flight. Whether the lack of a pinna really is an adaptation for flight is unclear, for the reptilian ancestors of birds had no pinna, so it is possible that the evolution of the pinna in mammals was an adaptation to improve hearing in a group that was primarily nocturnal. It is obvious that the presence of a pinna does not prevent flight, for many bat species have enormous external ears (yes, I know, they don't fly as swiftly as birds). The other way of looking at this is to consider the fact that none of the fifteen families of flightless birds has an external ear; nor did the most primitive birds have external ears. My guess, therefore, is that the lack of a pinna is a consequence of the birds' ancestry rather than an adaptation for flight.[16]

The value of our own pinna is all too apparent. By cupping our hand round our ear we increase the effective size of the pinna and the effect is dramatic. In much the same way, in recording birdsong (or anything else) a parabolic reflector on a microphone increases the amount of sound gathered. The lack of a pinna must potentially have a marked effect, not only on how well birds can hear, but also on their ability to pinpoint the source of a particular sound – although, as will become clear, birds have evolved other ways of doing this.

A *second* difference between birds and mammals is that mammals, including humans, have three tiny bones in the middle ear, whereas birds have only a single bone, as do reptiles, again reflecting their evolutionary history.[17]

Third is the inner ear – the business part of the ear. It is embedded in bone, for protection, and comprises the semi-circular canals (concerned with balance, which we won't discuss here) and the cochlea. In mammals the cochlea is a spiral structure (*cochlea* means snail in Latin), whereas in birds the cochlea is straight or slightly curved like a banana. Inside the fluid-filled cochlea lies a membrane – the basilar membrane – on which there are lots of tiny hair cells. Sensitive to any kind of vibration, the hair cells work like this. A sound occurs, producing a pressure wave which travels down the auditory canal in the external ear until it hits the eardrum. This now causes the bone(s) of the middle ear to vibrate, which in turn transmits a vibration to the beginning of the inner ear and then to the cochlea. A pressure wave occurs inside the fluid of the cochlea causing the hair of the hair cells to bend, firing off a signal to the brain. Sounds of different frequencies – which I'll explain in a moment – reach different parts of the cochlea, stimulating different hair cells. High-frequency sounds cause the base of the basilar membrane to vibrate, and low-frequency sounds cause the far end of the membrane to vibrate.

The coiling of the cochlea in mammals allows a greater length to be packed into a small space, and, indeed, the mammalian cochlea is longer than that of most birds: about seven millimetres in mice and just two millimetres long in the similarly sized canary. One possible explanation for this difference is that a coiled cochlea enhances the detection of the low-frequency sounds used by many large mammals.[18]

One of the pioneers of the avian inner ear was the extraordinarily talented Swedish scientist Gustav Retzius (1842–1919). By marrying Anna Hierta, the daughter of a newspaper magnate, Retzius gained financial independence and almost complete freedom to pursue his studies, which ranged from the design of spermatozoa to poetry and anthropology. It is his work on the nervous system and the structure of the inner ear, however, for which he is best known. Retzius was one of the first to provide comparative information and beautiful illustrations of the inner ear of a range of animal species, including several birds. Poor Reztius! Nominated no fewer than twelve times for a Nobel Prize, he never quite made it to Stockholm. When Jerry Pumphrey later took stock of what was known about the senses of birds in the 1940s, he put Retzius's detailed descriptions to good use, speculating about the hearing ability of birds by dividing them into those whose cochlea was 'conspicuously long' (eagle owl); long (thrushes and pigeons); average (lapwing, woodcock and nutcracker); short (chicken); and very short (goose, sea eagle). Pumphrey wrote: 'If we exclude the owl, we can perhaps imagine a correlation between the length of the cochlea and musical ability.' He was not far off. We now know, first, that the ears and hearing of owls differ from those of most other birds, and, second, that if we interpret 'musicality' as its reciprocal, 'the ability to detect and distinguish sounds', then Pumphrey's speculation is remarkably accurate.[19]

With the benefit of more information both on cochlea size and hearing ability, it is now apparent that the length of the cochlea (specifically, the basilar membrane inside it) is a reasonable index of a bird's sensitivity to sound. As with other organs (brain, heart, spleen), larger birds have a larger cochlea, but, in addition, larger birds are also particularly sensitive to low-frequency sounds, and small birds more sensitive to high-frequency sounds.

Let's put some numbers on this, so we can see the pattern – we'll use just five species: the zebra finch (which weighs around 15 g) has a basilar membrane about 1.6 mm long; budgerigar (40 g), 2.1 mm; pigeon (500 g), 3.1 mm; gannet (2.5 kg), 4.4 mm; and emu (60 kg), 5.5 mm. The existence of this relationship means that researchers can predict how sensitive a bird is to particular sounds from the length of its cochlea. Indeed, biologists have recently done just that, using the dimensions of the inner ear of the extinct Archaeopteryx – derived from fMRI imaging of the fossil skull – to suggest that its hearing was probably much like that of a present-day emu – that is, rather poor.[20]

Owls are the exceptions. For their body size, their cochlea is, relatively, enormous and contains very large numbers of hair cells. The barn owl, for example, which weighs around 370 g, has a relatively enormous basilar membrane at nine millimetres, containing some 16,300 hair cells – more than three times what we would expect from its body size, and providing exceptionally good hearing.

Fourth, the hair cells within the cochlea of birds are replaced on a regular basis, whereas those of mammals are not. Had the corncrake that called so close to my ear remained where it was and continued to call, and had I been silly enough to continue to lie there, the volume of its call would eventually have started to damage my ear and impair my hearing – irreparably. The hair cells responsible for detecting sound in the inner ear are so sophisticated and so delicate that they are easily damaged by too much noise. Ours is a sensitive system. It is so sensitive, in fact, that any further improvement and we would hear the sound of our own blood rushing through our heads. Rock musicians and their fans know to their cost the long-term damage to the ear caused by too much noise. Damaged hair cells are not replaced. This is also why, as we get older, we find it increasingly difficult to detect high-frequency sounds. Many birdwatchers over fifty that I know are oblivious to the goldcrest's high-pitched song, or in the Americas are unable to hear the songs of species like the black-throated green warbler and the blackburnian warbler. And it is not just ageing rockers: Gilbert White, author of *The Natural History of Selborne* (1789), at the relatively young age of fifty-four, bemoaned how: 'Frequent returns of deafness incommode me sadly, and half disqualify me for a naturalist.'[21]

Birds are different in that their hair cells *are* replaced. Birds also seem to be more tolerant of damage created by loud sounds than we are. This is currently an area of intense research, for if we can establish the mechanisms by which birds replace their hair cells, a cure for human deafness might be found. So far the prize is elusive but in their quest researchers have discovered a great deal about hearing, including its genetic basis.[22]

Fifth, imagine what it would be like if our ability to recognise voices on the telephone disappeared each winter. Inconvenient? Yes, given our lifestyle, but birds' hearing ability really does fluctuate throughout the year.

One of the most remarkable of all ornithological discoveries was the realisation that birds in temperate regions undergo enormous seasonal changes in their internal organs. The most obvious of these involves the gonads. In a male house sparrow, for example, during the winter the testes are tiny, no bigger than a pinhead, but during the breeding season they swell to the size of a baked bean. The human equivalent would be testes the size of apple pips outside the breeding season. Similar seasonal changes occur in females: the oviduct, which is a mere thread of tissue during the winter, becomes a massive, muscular egg-delivery tube during the breeding season.

These enormous effects are triggered by changes in day length, which stimulates the release of hormones from the brain, and in due course from the gonads themselves. The hormones in turn trigger the onset of song in the males. Perhaps the most far-reaching discovery relating to these changes was the finding in the 1970s that parts of the brain also varied in size across the year. This was totally unexpected because the conventional wisdom was that brain tissue and neurons were 'fixed' – what you were born with you had to make do with until the day you died. Exactly the same was thought to be true of birds. The realisation that this was not the case in birds revolutionised and reinvigorated research in neurobiology and song learning, because, among other things, it has the potential to provide a cure for neuro-degenerative diseases like Alzheimer's.

The centres in the avian brain that control the acquisition and delivery of song in male birds shrink at the end of the breeding season and grow again in the following spring. The brain is expensive to run – in humans it uses about ten times as much energy as any other organ – so, for birds, shutting down those parts not needed at certain times of the year is a sensible energy-saving tactic.

In temperate regions birds typically sing most in the spring; this is when males establish territories, which they defend via song, and when they acquire a partner, which they attract via song. A few temperate birds, like dippers and nuthatches, however, set up their territories in late winter and start singing earlier in the year. The hearing ability of songbirds is most sensitive at the time of year when song is most important.

This makes sense. If song is predominantly a springtime event, it follows that it might be advantageous if birds' hearing ability is enhanced at this time. Males, for example, need to be able to distinguish territorial neighbours from non-neighbours who will pose more of a threat, and females need to be able to distinguish between potential partners of different quality. Studies of three North American songbirds, the black-capped chickadee, the tufted titmouse and the white-breasted nuthatch,[23] show that seasonal changes occur in both sensitivity (the ability to detect sound) and processing (the ability to interpret those sounds). Jeff Lucas, who conducted this research, suggested thinking about this as if these three species were listening to an orchestra:

> Chickadees show a broad-band increase in processing in the breeding season, so the orchestra really would sound better to them in the breeding season. Tufted titmice show no change

in processing but they do show a change in sensitivity, so the orchestra wouldn't sound any better, but it would sound louder. White-breasted nuthatches show a narrow-band increase, increasing processing of 2 kHz tones. So for them, the orchestra would sound better when the orchestra was playing a C(7) or B(6), but the timbre of the instruments wouldn't be more enjoyable.

You might be surprised to learn that humans also experience predictable, regular changes in their hearing ability – or at least females do. Oestrogen is the key: when oestrogen levels are high, a man's voice sounds richer. The effect is so subtle that most women are unaware of it, but, even so, it may play a vital part in mate choice.[24]

The sounds birds make vary from the deep booming of a bittern to the high-pitched tinkle of goldcrests and kinglets. The frequency (or pitch) of sound is measured in hertz – the number of sound waves going by at any one time, usually expressed in terms of thousands of hertz, or kilohertz (kHz). A bittern's boom clocks in at around 200 cycles per second, or 200 hertz (Hz), or 0.2 kHz. In contrast, the goldcrest sings at a frequency of around 9 kHz. These two sounds cover pretty much the entire span of sound frequencies uttered by birds. A canary, a typical songbird, sings at a frequency of about 2 or 3 kHz. As we might expect, the frequency of the sounds birds make matches pretty closely what they can hear, or, to be more precise, the frequencies at which they are most sensitive. Humans hear best at about 4 kHz, but we can hear sounds as low as 20 Hz and as high as 20 kHz – when we are young. Birds are most sensitive to sounds in the region of 2 or 3 kHz and most are capable of hearing between 0.5 kHz and 6 kHz.[25]

What humans and birds can hear is usually illustrated by means of an 'audiogram' or 'audibility curve'. This is a visual representation of the quietest sound an animal can hear at different frequencies throughout its range of hearing. It comprises a plot of frequency (in kHz) along the horizontal axis and loudness up the vertical axis. The fact that the graphs are U-shaped indicates that for both birds and humans the quietest sounds we can hear are those in the middle frequency range; for us to detect lower or higher frequency sounds they need to be louder. The audiograms of humans and most birds are rather similar, although humans have better hearing at mid to low frequencies. Owls have better hearing than most other birds (and humans) in that they can detect much quieter sounds, and songbirds have better hearing at high frequencies than other birds. Although only a few species have been tested, it seems likely that bitterns are most sensitive to low-frequency sounds, and goldcrests to high-frequency sounds.

Birds use their hearing to detect potential predators, to find food and to identify members of their own and other species. To be able to do all these things they must be able to identify where a particular sound is coming from; distinguish meaningful sounds from 'background' noise created by other birds and the environment; and discriminate between similar sounds, much as we can recognise the voices of different people.

Imagine you are alone in the dark in an unfamiliar place, and unsure about how safe you are. Suddenly there's a strange sound, a footstep on gravel perhaps . . . but you cannot tell which direction it has come from. Is it behind you, in front, or off to one side? Knowing exactly where a potentially dangerous sound comes from is crucial if you are to prepare yourself for a quick escape. Being unable to localise a sound – especially in a dangerous situation – is one of the most disquieting of experiences. We are normally pretty good at localising sound and, of course, when it isn't dark we use our eyesight to check and confirm the source of sounds.

We pinpoint sound by unconsciously comparing when it reaches each of our ears. Our heads are large enough, and our ears far enough apart, for a sound to reach our ears at slightly different times. In cool, dry air at sea level sound travels at 340 m per second, and that means that the maximum time difference in a sound reaching our two ears is 0.5 millisecond (one millisecond is 1/1000th of a second). If we detect no difference in the time when sound reaches our ears we assume that the sound is coming from directly in front of (or directly behind) us. Birds' heads are smaller than ours, and some, like hummingbirds, goldcrests and kinglets, have particularly tiny heads, which means that, everything being equal, they would have difficulty localising sound. Indeed, with just one centimetre between the ears, the difference in time of arrival of sound to the two ears would be less than 35 μm seconds (one microsecond (μm) is 1/1,000,000th of a second). Small birds get around this problem in two ways: first, by moving their heads more than we do, effectively increasing their size, enabling them to detect time differences; and secondly, by comparing the tiny difference in the *volume* of noise reaching each ear.

The type of sound also influences how easy it is to identify its origin and birds have exploited this in the way they communicate. It has long been known that when birds like thrushes or chickadees spot a predator, such as a hawk, flying overhead, they utter a high-pitched 'seep' call. Their high frequency (8 kHz) may render these calls inaudible to the predator (given that most predators are larger than their prey and that larger birds hear higher frequency sounds less well). The structure of these warning calls, which start and end imperceptibly and thus make them especially hard to locate, is exactly what you might expect from a signal where the signaller does not want to draw attention to itself. In contrast, when those same species spot a roosting owl they utter a completely different type of call, characterised by a harsh, abrupt chattering: a much more easily located sound. That's the whole point. When songbirds discover a non-hunting predator, they want to attract attention to it, recruiting songbirds to join in the mobbing and help drive the predator away. One of the interesting aspects of these two types of call is that they sound very similar across a range of species.[26]

The great French naturalist Georges-Louis Leclerc, better known as the Comte de Buffon, wrote this about owls in his history of birds of the mid-1700s: '[their] sense of hearing . . . appears to be superior to that of other birds, and perhaps to that of every other animal; for the drum of the ear is proportionately larger than in the quadrupeds, and besides they can open and shut this organ at pleasure, a power possessed by no other animal'. Buffon is referring here to the enormous ear openings of certain owls, which in some species span almost the entire height of the skull, as my experience with a great grey owl demonstrated.

The great grey owl's greatness is partly an illusion; its enormous size is a consequence of its fabulous fluffy plumage. In reality, it is a midget in a vast, downy coat. The captive great grey owl whose ears I examined lay in the arms of his owner, looking up at me like a wide-eyed baby. As I carefully felt around behind his eyes, I couldn't believe the depth of the feathers and the smallness of its skull. Fully ten centimetres of feather created the bird's gigantic head. Around each eye the facial discs were bounded by a line of tawny feathers, conveniently marking the rear edge of the crevice in which each ear opening lay. Gently lifting the feathers forward on one side, the ear opening was revealed. It was huge – some four centimetres from top to bottom – and bewilderingly complex; the opening was covered by a moveable flap and bounded by unusual feathers. At the front edge, running from top and bottom, was a pallisade of rigid broad-shafted feathers, while the rear edge of the flap was lined with delicate filamentous feathers, behind which was a dense baffle of feathers that reminded me of a phalanx of Roman swords. The opening itself was huge, and contained lots of loose bits of skin, a bit like a grubby human ear. I then turned to the other side, and even though I knew that this species' ears are asymmetric, the degree of asymmetry astounded me. Looking at the bird face-on, the right ear lay below the level of the eye at seven o'clock; the left ear was at two o'clock. The bird's enormous head feathers are there simply to support the facial discs, gigantic reflectors whose purpose is to direct sound towards the ear openings.

One afternoon in the 1940s Clarence Tryon came across a great grey owl hunting in the Montana woods. The bird was perched on the end of a branch about four metres above the ground:

> Within a few minutes the owl made three swoops from its perch, apparently without catching anything. On the fourth swoop it hit the ground with considerable force and . . . flew away with a dead pocket gopher in its talons. The gopher was probably heard digging by the owl, which gave every indication of listening to some sound before swooping. Inspection of the spot . . . revealed that the owl had apparently broken through the thin roof of one of the feeding runways of the gopher's burrow.[27]

Subsequent observations by others revealed that great grey owls use the same technique to capture rodents beneath the snow – that is, entirely by sound:

> Watching and listening, the owl turns its head from side to side, occasionally peering intently towards the ground. Once the prey animal is detected, the owl dives down, appearing to hit the snow with its head but in fact at the last moment the feet are brought forward beneath the chin to grasp the prey.[28]

To be able to hunt by sound alone, great grey owls must have extraordinarily acute hearing, but they also need to be able to pinpoint the source of sound very accurately in both the horizontal and vertical plane. They do this through a remarkable suite of auditory adaptations which include the facial discs, each of which acts as a large pinna funnelling sound towards the inconspicuous ear openings. Early naturalists including John Ray and Francis Willughby commented in the 1670s on the fact that the barn owl's eyes were: 'sunk in the middle of [the facial feathers], as it were in the bottom of a pit or valley'. What Ray and Willughby did not appreciate was that the valleys on either side of the face, created by the facial disc, increase both the effectiveness with which sound is 'gathered' and the bird's ability to localise sound. Three centuries on, and with the benefit of much greater knowledge, Masakazu Konishi, studying the hearing of owls, wrote: 'When one sees the whole design of the facial disc, one cannot help thinking of a sound-collecting device.'[29]

A second adaptation – known since the Middle Ages – is the relatively enormous ear openings of species like the great grey owl. The term 'ear' is potentially confusing – some owls are 'eared' – the great horned, long-eared and short-eared owls have feathers on top of their heads that look superficially like ears, but have nothing to do with hearing. What I'm referring to are the genuine ear openings which, as in the great grey owl, are also asymmetrical, one being higher than the other. Many owl species possess asymmetric ear openings and in most instances this affects only the soft tissues of the external ear, but in the boreal (or Tengmalm's), saw-whet, Ural and great grey owl, the skull itself is also asymmetric, although the internal structure of each ear is identical.

The significance of this was recognised in the 1940s when Jerry Pumphrey pointed out that asymmetric ears would make it much easier for the owl to pinpoint the source of sound. In the 1960s, Roger Payne of the New York Zoological Society (and later famous for his studies of whale songs) conducted an ingenious experiment on a captive barn owl in a completely dark room, to demonstrate this. When the light was reduced over several successive days, the owl – which was observed by infra-red light (invisible to owls) – was able to catch mice in total darkness simply by homing in on the sound of the mice rustling leaves that covered the floor. As a test of what the owl was homing in on, Payne conducted an experiment in a room whose floor was covered with foam rubber, tying a dry, rustling leaf to the tail of a mouse. The owl swooped on the leaf (the source of the sound) rather than the mouse itself, dispelling the idea proposed earlier than owls might have infra-red vision or some other sense, confirming that sound alone was the cue.[30]

Interestingly, the owl was able to capture prey in total darkness only if it was thoroughly familiar with the layout of the room; a bird moved to a new room was reluctant to hunt in total darkness. This makes sense: swooping around with no light whatsoever is potentially extremely dangerous, unless, of course – like the oilbird, which we'll discuss in a moment – it has some additional sensory mechanism. It is also striking that, on capturing prey in total darkness, the owl instantly turned on itself in order to return directly to its perch, thereby avoiding any unnecessary flying around in the dark. The need to be familiar with the topography before hunting in total darkness explains why certain nocturnal owls remain in the same territory for most of their lives. Nights with no light whatsoever are few, but when they occur (for example, when there is heavy cloud cover and no moon) detailed topographical knowledge must determine whether or not an owl secures a meal without injuring itself.[31]

One of the most intriguing features of owls is their extremely quiet flight; their wingbeats are almost imperceptible. When Masakazu Konishi analysed the sound of one of his barn owl's wingbeats, he was surprised by its low frequency – around one kilohertz. The beauty of this is that, even when the owl is flying, it does not interfere with its ability to hear its prey. The rustling of mice in the undergrowth has a much higher frequency of between six and nine kilohertz. What's more, since mice are relatively insensitive to sounds lower than about three kilohertz, they are unable to hear an approaching owl.[32]

I return to Skomer Island each summer to continue the study of guillemots I started during the 1970s. The high spot of the season is climbing on to the breeding ledges to catch several hundred chicks to ring (band) them so that we can later assess how old they are when they start breeding and how long they live. Ringing involves climbing on to the breeding ledges and using a carbon-fibre fishing pole tipped with a shepherd's crook to catch the chicks. This is a social event, involving one catcher; one taker (who removes the chicks from the crook and places them in a net bag, prior to ringing); one ringer; and a 'scribe' (the note-taker, who records which ring is placed on which bird). It is also a very noisy affair, for the parent guillemots temporarily deprived of the chicks call loudly, and, deprived of their parents, the offspring reply in a much higher register. It is sometimes so noisy on the ledges that we have to shout the ring numbers to the scribe. By the end of a day's ringing, our ears are often ringing too.

The chicks themselves are very good at distinguishing their parents, and vice versa. In fact, they have learned each other's calls from before the chick even emerged from the shell: as soon as the first hole appears in the egg, the chick and adult start calling to each other. Under normal circumstances a guillemot colony is fairly noisy, but chicks stick close to their parents and there's usually no great need to maintain continuous vocal contact. But if a gull or other predator causes adults to temporarily abandon their chick, it is vital that parents and chicks can find each other as soon as they return. This is especially important when it is time for the young guillemots to leave the colony, which they do en masse at about three weeks of age, at dusk. The chick, which is still flightless, usually jumps off the ledge into the sea below, either to join its father waiting on the sea or to be swiftly followed by its father. Staying together is vital. In very large colonies, like the one on Funk Island off the coast of Newfoundland, there may be tens of thousands of young birds leaving on the same night, and it can be difficult for fathers and chicks to remain in contact. They do so by means of their individually distinct calls. The departure of guillemot chicks is a cacophony of calling: a high-pitched *weelow weelow weelow* from the chick and a harsh, guttural growl from the adult. Remarkably, the vast majority of adults and chicks find each other on the water and swim off together out to sea where they remain, still together, for a further few weeks.

The guillemots' hearing is so good that they can cut through the wall of noise, and single out those calls that really matter. This is literally a life and death situation, for unaccompanied chicks die. Natural selection has produced a hearing system that enables both adult and young guillemots not just to hear each others' cries, but to distinguish them from all the extraneous cries around them. The way birds do this is by filtering and ignoring irrelevant noise, focusing instead only on those sounds that matter both for identifying their own species, and also for identifying specific individuals.

The ability to focus on one particular voice or song against a hubbub of background noise is known as the cocktail-party effect. This is a common problem for birds that live in a noisy world. Just think of the dawn chorus: in pristine habitats there may be as many as thirty different songbird species – with several individuals of each – singing at once, and the effect can be almost deafening. Each bird has to distinguish not only its own species, but also different individuals. In much the same way, starlings coming in to roost in city centres often stop off on a church tower or other high structure and start singing in their hundreds. Can they really distinguish each other in such swarms? The answer is probably 'yes'. In some experiments (using relatively modest numbers compared with the huge roosts one often sees), captive starlings were able to discriminate between individuals on the basis of song even when the songs of several other starlings were played at the same time.[33]

As well as having to deal with the sounds of other birds, the physical environment has a huge effect on what birds can hear. For marine birds it is the sound of the sea crashing against the cliffs at the colony; for birds breeding in reedbeds it is the rustling of a profusion of reeds; for those in rainforest it is rain falling on a million leaves.

It is obvious – and has been known for a long time – that sound gets fainter with distance. The term used to describe this degradation is 'attenuation' and it is also well known that attenuation differs between habitats. In a flat, open habitat sound travels further than in a forest or a reedbed. The first studies of the effect of attenuation on birdsong in different habitats were conducted in the 1970s. Their results had been anticipated, albeit unconsciously, by the makers of the 1940s Tarzan movies whose soundtracks often featured very specific bird calls – calls that we still associate with rainforest habitat: low-frequency, extended flute-like whistles. Gene Morton, based at a biological station, the Smithsonian Tropical Research Institute, in Panama, noticed the same thing, and wondered whether such calls had been shaped by natural selection for better transmission in dense habitats. The key to establishing whether the quality of the sound affects how well it can be heard at a distance was first to measure the attenuation of different-quality sounds in different habitats. Morton did this by playing sounds from a tape recorder and measuring their quality at different distances and in different habitats. Having shown that low-frequency, pure sounds travelled further in rainforest than other types of calls, Morton then recorded birds from both forest and adjacent open habitat and compared their calls. As he predicted, the forest-dwelling birds had lower frequency calls. In general, low-pitch calls travel further than high-pitch ones, which is why foghorns use deep sounds, and why bitterns and kakapos are among the record-holders for sound transmission.[34]

Morton's study was based on a comparison of different bird species, but other ornithologists wondered whether it might also be true within a single species living in different habitats. One of the first single-species studies was undertaken by Fernando Nottebohm on a very common and widespread Central and South American bird, the rufous-collared sparrow, known locally at as the chingolo. As predicted by Morton's cross-species comparison, the chingolos' song contained more long, slow whistles in forested habitat and more trills in open habitat.[35] Similar results were later obtained for Eurasian great tits breeding in dense forest compared to more open woodland habitat.[36]

Dramatic evidence that birds respond appropriately to background noise comes from recent studies of birds in urban environments. Nightingales in Berlin sing louder (by a huge 14 dB) than their rural counterparts, and sing more loudly on weekday mornings during rush hour, when the traffic noise is loudest. Great tits, on the other hand, do not change the volume of their songs, but change their frequency or pitch to cope with urban noise. In both species the birds adjust their singing behaviour to ensure that they can still be heard, despite the background noise.[37]

Increasing the volume of sounds uttered in a noisy environment is actually a reflex known as the Lombard effect, named after Etienne Lombard, a French ear, nose and throat specialist who discovered it in humans in the early 1900s. The Lombard effect is most obvious when somebody is talking to you when, for example, you have your iPod headphones on and in response you – unwittingly – increase the volume of your reply, and they say: 'No need to shout!'

I visited New Zealand while writing this book, and when I wasn't chasing kiwi and kakapo I took a few days off to visit Fiordland on South Island. The weather was perfect and the scenery spectacular, but the most striking aspect of this area was its auditory desolation. I have rarely been anywhere so quiet. Peaceful, yes, but this was a melancholic silence. The birds that once inhabited the forests clothing the steep-sided valleys have all been killed by the predatory stoats and weasels that the early settlers foolishly introduced. Native birdsong is absent across mainland New Zealand and made me wonder whether the introduced dunnocks, blackbirds and thrushes sing more softly in New Zealand than in their native Europe due to the lack of competition.

The studies I have just described show clearly that habitat affects the types of songs birds use, and in a way consistent with what is known about sound attenuation. These studies, however, provide only indirect evidence that the birds themselves hear sounds differently in different habitats. Some nice evidence that they do comes from a study of the North American Carolina wren which uses song to defend its territory throughout much of the year. The presence or absence of leaves on the vegetation (in winter and summer, respectively) has a huge impact on the way songs sound. The wrens' songs degrade over distance more rapidly when there are leaves on the vegetation than in winter when there are no leaves. When Marc Naguib broadcast undegraded or degraded songs at the same volume and from the same location, the wrens typically responded to the undegraded song by flying directly to the loudspeaker. When he broadcast degraded songs, however, the birds flew *over* the loudspeaker, as though they perceived the intruder to be further away. In other words, the wrens could tell the difference between degraded and undegraded song and adjusted their behaviour accordingly.[38]

The audio equivalent of a microscope or high-speed camera is the sonograph, a machine that produces a picture of sound. Invented during the 1940s by the Bell Telephone Laboratories in the United States, the sonograph was first used by W. (Bill) H. Thorpe in Cambridge to understand birdsong. Being able to 'see' sound, as a sonogram, transformed the study of birdsong. Of course, tape recorders had been around before this, but listening to a bird's song, even if played at reduced speed, doesn't really provide the same resolution or sense of understanding as an image. Only by transforming an auditory signal into a visual one did we really start to appreciate the full complexity of birdsong, and to speculate about how much of that complexity the bird actually hears, or can make sense of. As an undergraduate I completed a three-month-long project on the contact calls of golden-breasted waxbills and can still remember the distinctive acrid smell of the sonograph machine as it burned the sound image (the sonogram) on to the heat-sensitive paper.

If one listens to the song of the whippoorwill – a North American nightjar – it sounds, as its name implies, as though it consists of three notes, rendered in David Sibley's *Guide to Birds* as: WHIP puwiw WEEW (i.e. whip-poor-will). If one then makes a sonogram of the call, it becomes clear – from this 'slow-motion' visualisation – that the call actually consists of *five* separate notes, not three. To the human ear the call is delivered so fast that the separation of the individual notes is obscured. When this was discovered by the ornithologist Hudson Ansley in the 1950s it was not clear whether the whippoorwill itself hears three or five notes, for at that time little was known about birds' hearing. However, as Ansley pointed out, if one looks at a sonogram of a mockingbird imitating a whippoorwill, it uses five rather than three notes, suggesting that this species at least can resolve the fine detail of the whippoorwill's song.[39]

Tests of human hearing indicate that our ability to resolve different sounds starts to break down as the interval between the sounds approaches one tenth of a second. Many birds' songs, however, contain elements occurring at much shorter intervals than this and there is increasing evidence that birds are able to detect such differences. Indeed, this is the one aspect of hearing in which birds are much better than humans. It is as if they have the auditory equivalent of a slow-motion option in their brain, allowing them to hear details that are completely lost on us. This raises an interesting point: if we were able to hear birdsong exactly as a bird hears it, would we still consider it 'beautiful'; would we still consider birdsong to be akin to music?

Striking evidence of the ability of birds to hear the fine details of song involves the so-called 'sexy syllables' in canary song. When a male canary sings in front of a female around the time she is laying eggs, her response is often to solicit copulation by crouching. Detailed analysis reveals that the part of the song that triggers this response is a succession of rapidly alternating high- and low-frequency elements (produced from the right and left sides of the syrinx – the bird's voice box – respectively) at a rate of about seventeen times per second. To us, the burst of sexy syllables midway through the song sounds like a continuous trill, but the females hear the fine detail. Using a computer to create artificial songs, Eric Vallett manipulated different components of the sexy syllables, making them faster or slower by altering the interval between the syllables and then playing them to females. The female canaries had no difficulty distinguishing between the two songs, demonstrating their preference for the faster trill by crouching for copulation.[40]

Driving through Ecuador's monumental mountain scenery, we start to descend into a forested valley on a road so steep it feels like zooming in on Google Earth. Down, down, down, slithering and sliding on the rough track until, after forty-five minutes, we eventually stop in a cloud of dust beside a small ravine. It doesn't look very promising: a crudely constructed bamboo scaffold supports a black plastic pipe emerging from a cleft in the rocks. Treading over plastic rubbish, boulders and dead leaves, we make our way gingerly up the sunless gorge. Within a few metres we turn a corner and are suddenly confronted by three oilbirds sitting on a low, muddy shelf. They are just as shocked by our intrusion as we are by their proximity. Without warning, they clatter into the air shrieking and clicking like demons. In fact, that's just what they seem like, medieval birds more fitting for a Harry Potter film than the tropics. Their local name is *guácharos* – literally 'one who cries and laments', and is possibly onomatopoeic . . . likened by others to the sound of ripping silk. Their scientific name, *Steatornis*, which literally means 'oil bird', refers to the fact that in the past their very fatty chicks were rendered down for oil which was used in cooking.

The birds settle eventually on a ledge ten metres up and sit in close bodily contact. Like a cross between a hawk and a nightjar in appearance, 'nighthawk' might have been a good name for them, although they are far from hawk-like in their habits. They have huge, dark eyes; a walrus-moustache comprising twelve long bristles sweeping down from each corner of the mouth; an enormous hawk-like bill with distinctive oval nostrils; and perhaps most striking of all, rows of brilliant white spots decorating their russet plumage. There are three rows of spots, running along the wings, the tail and on the breast, and more on the top of their head – like a scattering of stardust. We stand stock still, rooted to the spot, partly in awe and partly in fear of disturbing these extraordinary birds. After fifteen minutes they seem to relax, closing their eyes and returning to the sleep from which we disturbed them. As our own eyes become accustomed to the gloom, and as theirs adjust to the light, we see more and more birds distributed on ledges and in small caves. Our guide tells us that there are about a hundred in total: all the more remarkable because this might be one of the few places in Ecuador where oilbirds live. But the birds are desperately vulnerable. The plastic water pipe running through the gorge comes from a newly built road only tens of metres above the birds.

Construction of this road is a vicious rent through the forested valley bottom, stretching wider and wider as it extends across the landscape, stripping away the forest on either side. Once the road is open I wonder how long the *guácharos* will last; it is hard to imagine them dozing through the day with a barrage of noisy trucks thundering overhead in a haze of diesel fumes. It is hard to imagine, too, when the trees have gone, how they'll be able find enough fruit.

The oilbird is one of just a handful of birds that relies – like many bats – on hearing the echoes of its own voice to navigate in total darkness. It is well known that bats use echolocation to operate in the dark, but that particular discovery was protracted and hard-won.

The pioneer of bat senses, and much else, was Lazzaro Spallanzani (1722–99), Jesuit priest and professor of natural sciences at the University of Pavia, Italy. Endlessly curious about the natural world, Spallanzani was a brilliant observer and ingenious experimenter. Watching a captive barn owl he noticed that if the bird accidentally extinguished the candle that was lighting the room, the bird lost all ability to avoid collisions. Bats had no such problem. Placed in total darkness, the bats Spallanzani collected from a local cave 'continued to fly around as before and never struck against obstacles, nor did they fall down as would have happened with a night bird [i.e. the owl]'. Two bats whose eyes Spallanzani covered with a dark hood also flew quite normally.

These phenomena induced me to perform another experiment which I considered decisive, namely to remove the eyes of a bat. Thus with a pair of scissors I removed completely the

eye-balls in a bat . . . Thrown in the air the animal flew
quickly, following the different subterranean pathways from
one end to the other with the speed and sureness of an unin-
jured bat . . . My astonishment at this bat which absolutely
could not see although deprived of its eyes is inexpressible.[41]

Spallanzani wondered whether the bats possessed a sixth sense. He
wrote to everyone who could help, offering a challenge: could
anyone discover how blinded bats could 'see' in the dark? One of
Spallanzani's letters was read to the Geneva Natural History Society
in September 1793, where the Swiss surgeon and natural historian
Charles Jurine was in the audience. Intrigued, Jurine decided to
conduct his own experiments and began by repeating what
Spallanzani had done, but adding an ingenious twist. As well as
removing their eyes, he also plugged the bats' ears with wax, and to
his amazement found that they 'blundered helplessly into all obstacles'.[42] The conclusion was extraordinary: bats needed to hear to be
able to 'see'.

Spallanzani learned of Jurine's remarkable results the very next
day and immediately started some new experiments of his own,
deafening bats and confirming that they rely on reflected sound,
but with no notion of where it came from. Puzzled, he said: 'But
how, if God loves me, can we explain or even conceive in this
hypothesis of hearing?' Given that the bats were silent, why were
their ears so important in avoiding obstacles? The experiments gave
the same results, time after time; the problem was that, unable to
imagine that certain sounds might lie outside the range of human
hearing, they simply did not make sense.

Georges Cuvier (1769–1832), the renowned and influential
French anatomist, decided in 1795, on the basis of little more than
logic, that bats avoided obstacles through a sense of touch. Even
though Spallanzani had earlier tested and comprehensively rejected

the touch hypothesis, Cuvier's idea became the accepted explanation and he was 'lauded for having brought order out of a chaotic
state of affairs left by Spallanzani and Jurine'. The reason why
Cuvier carried the day was that, with no notion that bats might
utter sounds inaudible to humans, Spallanzani's and Jurine's ideas
seemed completely fanciful.[43]

The touch idea remained unchallenged for one hundred years,
by which time two further possibilities were in the air. The first
surfaced after the sinking of the *Titanic* in April 1912. Impressed by
the ability of blinded bats to avoid collisions, the engineer and
inventor Sir Hiram Maxim wondered whether ships might similarly be protected from collisions with icebergs and other ships in
foggy weather by an apparatus that could detect the returning
echoes from powerful low-frequency sounds. He assumed that bats
heard and responded to the reflections of the *low*-frequency sounds
made by their wingbeats. In other words, Maxim was the first to
suggest that bats might use sounds inaudible to the human ear.

The second idea was the brainchild of physiologist and sound
specialist Hamilton Hartridge (1886–1976), who was reminded of
the underwater object detection techniques developed during the
First World War. He wondered whether bats avoided obstacles by
the reflections of what he assumed were their *high*-pitched calls.

Of the two ideas, Hartridge's high-frequency sounds seemed the
more plausible and in the early 1930s Harvard undergraduate Don
Griffin decided to test it. He did so using the only piece of kit
capable of detecting and analysing high-frequency sounds: elec-
tronic equipment constructed by the physicist George Pierce to
detect the high-frequency sounds made by insects. It is not unusual
for researchers to design and build their own research equipment,
and Griffin was fortunate that Pierce was happy to share his technology. The outcome was remarkable, and confirmed beautifully
that bats utter sounds beyond the range of normal human hearing.

Most people can hear sounds with frequencies as low as 20 or 30 Hz, and as high as 20 kHz, but the bats that Griffin studied were uttering cries as high as 120 kHz.[44]

Together with a fellow student, Robert Galambos, Griffin then began more detailed investigations. Their efforts in the early 1940s resulted in the momentous discovery that bats not only utter a continuous stream of high-frequency sounds, but they do so at an increasing rate whenever they are negotiating particularly tricky objects. This provided strong circumstantial evidence for Hartridge's idea that bats avoided obstacles using the echoes from their high-pitched cries. Coincidentally, it was also realised around this time that visually impaired people could detect obstacles by making sounds and hearing the reflections of these sounds, inspiring Griffin to coin the term 'echolocation' for the process. Ten years later, Griffin was able to show that, as well as using echolocation to avoid obstacles, bats also use it to hunt their insect prey. This, too, was totally unexpected. The conventional wisdom before he started was that tiny flying insects would not 'return enough acoustical energy to yield audible echoes, and the whole idea seemed too far-fetched for serious consideration'.[45] Yet this is exactly what he found, confirming that the bat's echolocation system was far more sophisticated than anyone had imagined.

Excited by his discoveries, Griffin next went in pursuit of oilbirds, to check whether they also used echolocation to orientate themselves in total darkness. In 1799 – the year Spallanzani died – the German naturalist and explorer Alexander von Humboldt was in tropical America with a botanical colleague, Amie Bonpland. At Caripe in Venezeula they visited the Guácharo Cave, an enormous cavern inhabited by thousands of nocturnal birds, which the local people were very reluctant to enter. As Humboldt says: 'The cave at Caripe is the Tartarus of the Greeks, and the Guácharos which hover above the torrent, emitting plaintive cries, recall the Stygian birds.'[46]

Humboldt named the bird *Steatornis caripensis* – the oilbird of
Caripe – and although he was impressed by the tremendous noise
the birds made as they flew around in the cave, he did not comment
on their ability to navigate in total darkness.

It wasn't until the ornithologist William (Billy) H. Phelps Jr of
Caracas got someone to expose film in Humboldt's cave (now
known as *Cueva del Guácharo*) in 1951, that there was evidence that
the darkness in there was complete, and that the birds must be able
to navigate in absolute darkness. Accompanied by Phelps, Griffin
went to the Caripe cave to see for himself. Unlike Humboldt, who
had endured a difficult climb to reach the cave, by 1953 it was poised
to become a major tourist attraction and Griffin was able to drive
directly to the entrance, where he was greeted by the cave's custo-
dian and guides. At that date the young birds were still being
harvested for their fat, although not to the same extent as in
Humboldt's day when thousands were taken.

As Griffin's party, consisting of Phelps and his wife, Kathy, Mr
and Mrs McCurdy and Mr Zuloaga and his son, entered the cave,
they walked past the oilbirds nesting in what he called 'the twilight
zone', for their main objective was to establish the degree of darkness in which the birds could fly. In the deepest part of the cave
– the part Humboldt's local guide had refused to enter – Griffin's
party turned off their flashlights and sat in the dark to allow their
eyes to adjust while the oilbirds circled noisily but invisibly 75 feet
above them. After twenty-five minutes everyone agreed that there
was absolutely no light this deep into the cave, a fact confirmed by
Griffin's film, which was exposed for a full nine minutes. 'Our first
question was thus conclusively answered; the guácharos did fly in
total darkness . . .' Nor were they silent: 'Our ears were bombarded
almost constantly by a variety of squawks, screeches, clucks, clicks
and shrieks . . . But whether these weird cries of the guácharos were
used for orientation was still uncertain.'[47]

Griffin and his colleagues made their way back towards the cave entrance, and, as they did so, something remarkable happened. Outside, darkness was falling and the birds were beginning to leave the cave in search of fruit to feed their chicks. As the birds streamed out towards the cave entrance, instead of the ear-piercing calls they had uttered earlier, their calls were completely different: 'a steady stream of the sharpest imaginable clicks'. Subsequent analysis confirmed that these clicks had a frequency within the range of human hearing and much lower than most of the bats Griffin was familiar with.[48]

The next question was whether the oilbirds were using these audible clicks to navigate in the dark. An experiment was necessary. With some difficulty Mr Phelps and the local guides caught some birds by stringing a net across the cave entrance, and Mr Zuloaga arranged for Griffin to use the laundry room at the Creole Petroleum Corporation, where he worked, for the experiment. The room, from which all light was excluded, measured about 12 ft square by 8 ft high (3.6 x 2.4 m) and the birds flew around this confined space without touching the walls. In the dark Griffin could hear their wingbeats and, of course, their clicking. However, he noticed that the birds were unable to avoid the cord from the electric light that hung from the ceiling, raising the question of whether they could detect something this small in nature.

The experiment consisted of blocking the birds' ears with cotton wool, which they sealed with glue. If the birds relied on echolocation to orientate themselves, hearing would be essential. Taking the three strongest birds, Griffin duly plugged their ears, and waited a few minutes for the glue to set. The birds were released into the darkened room. The results were spectacular. In each case the birds clicked vigorously, but immediately flew into the walls. On the removal of their earplugs, the birds' ability to avoid the walls was restored. When the light was on the birds avoided the walls, but also uttered far fewer clicks, suggesting that when there was enough light the birds relied mainly on their eyesight.[49]

Overall, even though based on just a few individuals, Griffin's simple experiments demonstrated convincingly that oilbirds use echolocation like bats. They also showed that, unlike bats that typically use high-frequency sounds barely audible to the human ear, oilbirds used a low-frequency sound.

These extraordinary results were later confirmed in the 1970s by Masakazu Konishi and Eric Knudsen by showing that the oilbirds' clicks had a frequency of two kilohertz, which coincided exactly with the most sensitive region of their hearing. Taking this result together with what was known about echolocation in bats, Konishi and Knudsen suggested that the oilbird's echolocation might be fairly crude and limited to detecting relatively large objects. Bats use very high-frequency sound, but also project that sound in a narrow beam, whose echo they then detect using their very sensitive ears, enabling them to detect very small objects and even moths in flight. Konishi and Knudsen tested their idea by placing obstacles (plastic discs) of different sizes in a narrow part of the oilbirds' completely dark cave, knowing that to pass this point the birds would have to detect these obstacles. Observing the birds, using infra-red light, they watched the birds blunder into discs less than 20 cm in diameter as though they did not exist. With larger discs the birds had no problem avoiding them.[50]

One other group of birds relies on echolocation: the cave swiftlets of South East Asia. Like the oilbird, these birds breed in total darkness deep inside caves, but unlike the oilbird their nests are constructed of the birds' dried saliva (and harvested for bird's nest soup). Writing in 1925, G. L. Tichelman described a two-hour canoe journey inside a cave in Borneo: 'during the whole time one travels through a thick rain of the birds' twittering. Countless swifts fluttered around close to the canoe. On the dirty white rocks in places numberless swift nests were built so close together that they resembled clusters of black pickles.'[51]

The American ornithologist Dillon Ripley described another swiftlet cave in Singapore:

> The entrance consists of two relatively narrow semicircular openings through which the birds dash without appearing to slacken speed. As they fly by they make a rending sound, like the tearing of silk. An observer who stands by the entrance will have birds pass within a foot or so of him, and the noise of their flight is a thrilling sound . . . It seems fairly clear that the clicking is a sonic device to prevent the birds dashing themselves into the walls of the cave; they seem not in the least to slacken speed as they dash into the darkness.[52]

Later, using similar experiments to those performed with oilbirds, Alvin Novick confirmed that in total darkness cave swiftlets – like oilbirds – use low-frequency sounds to navigate via echolocation.[53]

As Jerry Pumphrey pointed out, compared with the high-frequency sounds used by bats: 'The practical disadvantages of employing . . . low frequencies for echo-location are so considerable as to suggest that the bird's ear is incapable of being readily modified in the direction of increasing sensitivity to ultra-sonic frequencies.'[54]

Overall, the sense of hearing in most birds is fairly similar to our own, with the notable exception of nocturnal species and those that hunt and navigate by sound, such as owls, the oilbird and cave swiftlets. For me, however, the bird that best captures the extreme sophistication of avian hearing is the great grey owl. Its ability to pinpoint a mouse, invisible under the snow, by means of asymmetric ears, leaves me speechless.

3
Touch

A mallard duck dabbling in mud. Thumbnails show (*left*) the inside of the upper bill showing the tips of the touch receptors in the rim of the bill, and (*right*) a single touch receptor (*enlarged*) with its two types of nerve endings: Grandry (*small*) and Herbst (*large*) corpuscles – pale spheres.

In birds . . . the horny beak appears unlikely to be a suitable vehicle for a refined sense of touch . . . the presence of end-organs [nerve endings] . . . suggest that it is in fact the part of birds which is tactually the most sensitive.

Jerry Pumphrey, 1948, 'The sense organs of birds', *The Ibis*, 90, 171–99

For several years while my children were growing up, we had a pet zebra finch named Billie. Born blind, Billie thrived on human company and was particularly fond of my daughter, Laurie, who had reared him from a chick. He knew her voice, but more impressively he recognised her footstep, although how he did this was a mystery, for Laurie is an identical twin and Billie never became excited at the sound of her sister's footfall. On hearing Laurie's approach, Billie would burst into song, and would do so again as soon as she opened his cage door and he hopped on to her finger. After his initial excitement Billie would solicit Laurie to preen his neck, tipping his head to one side and raising the feathers on the back of his neck, adopting exactly the same posture as he would when inviting a zebra finch partner to preen him.[1]

Ornithologists refer to one individual preening another as allo-preening ('allo' meaning 'other'), to distinguish it from the more usual self-preening. If you have ever tried to allopreen a bird like a zebra finch, whose entire body is smaller than your thumb, a finger seems far too large and clumsy. My daughter, who has small hands, was able to perform something akin to allopreening by using her index finger and Billie loved it, keeping his eyes closed and occa-sionally twisting his neck as if to provide access to new areas, much like a human having his or her neck or back scratched. When I tried preening Billie, I was aware of how huge my finger seemed and how careful I had to be to ensure that I tickled rather than pummelled him. If I lost control and was a little clumsy, he'd snap out of his reverie and either peck me or move away.

As far as I could tell, Billie thoroughly enjoyed the sensation of being allopreened and the same seems to be true when male and female zebra finch pair members allopreen each other. While it is easy to infer that the recipient enjoys the sensation of being allopreened, it is rather more difficult to decide what the bird performing the allopreening experiences.

When I allopreened Billie's neck, I was acutely aware of the sensation of my fingertip on his skin and feathers, and I used that information to regulate the tiny amount of pressure I was applying. When zebra finches allopreen each other, does the preener have similar feedback?

At first glance a bird's hard and horny beak seems decidedly insensitive. To see what it would be like to allopreen with an inanimate beak, I sometimes preened Billie with a dried grass stem, which was even finer than a zebra finch beak. In fact, the grass stem was not as inanimate as I imagined since I could feel the sensation of touch transmitted through it and into my fingers. What's more, Billie quite liked being preened in this more focused way.[2]

The truth is that a bird's beak is far from inanimate. Tucked away in tiny pits in different parts of the beak (and the tongue) are numerous touch receptors, and it is these that enable the zebra finch and other species to fine-tune their allopreening.[3]

Touch receptors – in human fingers – were first discovered in the 1700s[4] but not in the beaks of birds until 1860, when they were found in those of parrots and a handful of other birds.[5] Given the nature of their beak, parrots seem an unlikely bird to have a touch-sensitive bill tip, but they do, and this nicely explains their marvellous dexterity.

The bill-tip organ was discovered by the French anatomist D. E.

Goujon in 1869. In fact, he found that all the parrots he looked at, including the budgerigar, possessed this organ which consists of a series of pits in the upper and lower beak, full of touch-sensitive cells. Goujon's brief account is wonderfully enthusiastic: 'It is . . . not enough to know the exact topography of an organ, it is necessary to penetrate its very substance and to divine its fundamental elements where possible', and this is exactly what he did with touch receptors.[6]

In terms of keeping your fingers intact, if you want to examine a bird's bill-tip organ, a duck is a much safer option than a parrot. When I first saw a drawing of the nerves in a duck's beak,[7] I was reminded of an experience I had as a zoology undergraduate in the late 1960s, when one of my favourite books was Ralph Buchsbaum's *Animals Without Backbones*, first published in 1938. Buchsbaum brought the biology of invertebrates to life in an extraordinary and compelling way. One chapter begins: 'If all the matter of the universe except nematodes [threadworms] were swept away, our world would still be dimly recognizable . . .'[8] In exactly the same way, if all the matter of a duck's beak apart from the nerves was swept away, the beak would be clearly recognisable. Simply seeing that remarkable network of nervous tissue left no doubt in my mind that the avian beak, far from being an inanimate tool, must, in some species at least, be a highly sensitive structure.[9] The remarkable arrangement of nerves in the duck's beak was discovered by the English clergyman John Clayton, Rector of Crofton, in the late 1600s, who wrote:

Dr Moulin and myself when we made our anatomies together when I was at London, we shewed to the Royal Society that all flat-billed birds that groped for their meat [food] have three pair of nerves that came down their bills; whereby as we conceived they had that accuracy to distinguish what was proper for food and what to be rejected by their taste when they did not see it; and as this was the most evident in a duck's bill and head, I drawed a cut [i.e. an illustration] thereof and left it in your custody.[10]

Effectively, what John Clayton is saying is this: imagine being given a bowl of muesli and milk to which has been added a handful of fine gravel. How good would you be at swallowing only the edible bits? Hopeless, I suggest, yet this is precisely what ducks can do.

To understand how this is possible, first catch a duck. Then turn it over and open its beak so that you can examine its palate. The most striking feature is a series of grooves radiating around the curved tip, but you need to look beyond these at the outer edge of the bill. What you should be able to see now is a series of tiny holes or pores – some thirty of them. If you look on the lower jaw, you will find even more – about 180. Examining these pores with a magnifying glass, you will see that from each one protrudes the pointed tip of a cone-shaped structure called a 'papilla', inside which is a cluster of around twenty to thirty microscopic sensory nerve endings – these are the touch receptors – that connect to the brain via that network of nerves.

Nineteenth-century German anatomists were the first to see touch receptors in the duck's bill-tip organ. There are two types. The larger and more sophisticated ones were discovered by, and named after, Emil Friedrich Gustav Herbst (1803–93), who found them first in bone in 1848, then on the bird's palate in 1849, then in skin in 1850 and on the bird's tongue in 1851. Herbst corpuscles, which are sensitive to pressure and hence touch, are oval-shaped structures about 150 μm in length and 120 μm wide (one μm is 1/1000th of a millimetre), but occasionally up to one millimetre long. The second type, Grandry corpuscles, named after M. Grandry, a Belgian biologist who first found them in 1869, are smaller (about 50 μm long and 50 μm wide) and simpler in design, and are sensitive to movement. The two types lie together in the cone-shaped body of the papilla, with the smaller Grandry corpuscles positioned over the Herbst corpuscles – in a most beautiful structure.

Elsewhere in the duck's beak, both inside and out, there are large numbers of Herbst and Grandry corpuscles, particularly towards the tip and edges of the bill, but not bundled up together as they are in papilla in the bill-tip organ. Indeed, in just one square millimetre of a mallard's bill there are several hundred receptors, all designed to pick up information about things in contact with the bill and what is inside the bird's mouth.[11]

When we watch a duck dabbling in muddy water at the edge of a pond, rapidly opening and closing its beak, it is straining food items from the mud, retaining what's edible and rejecting the mud, gravel and water. It does this very quickly and without being able to see what it is doing, relying on its sensitive bill-tip organ together with the other touch receptors scattered throughout the mouth, and, as we'll see in the next chapter, its taste buds. We simply do not have the sensory (or mechanical) apparatus to do the same, which is why we would fail the muesli and gravel test. Ducks do, of course, use their eyes when they forage but in a different way – for example, when they take a piece of bread out of your child's hand; but as the bread is grasped, its texture is detected by the bill-tip organ, and then, if it tastes okay, it is swallowed.

How does a zebra finch manage to allopreen its partner with such sensitivity? Like that of the parrot and duck, the tip of the zebra finch's beak is also packed with nerve endings.[12] There are also a lot of touch receptors inside the mouth and on the tongue, whose main function is to facilitate the husking of seeds on which the zebra finch lives, and which is accomplished by sophisticated manipulation of the seed between the tongue and the upper mandible.[13] But these same touch sensors are also responsible for converting mechanical sensations into nerve impulses and the feedback from them allows the preener to control how much pressure it applies.

There is an apparent contradiction here: on the one hand I'm saying that the bird's beak is much more sensitive than is generally supposed, but on the other you may be wondering about woodpeckers using their bills as an axe. How can a beak be simultaneously sensitive and insensitive? The answer is: our hands work in exactly the same way. Formed as fists, our hands become weapons, but opened flat they are capable of the most sophisticated sensitivity – exemplified by Wilder Penfield's hugely handed homunculus.* A woodpecker hacks wood using the sharp, insensitive tip of its beak; it doesn't use the much more sensitive inside of its mouth. My concern is for those wading birds like the woodcock and kiwi whose bill tip is relatively soft and incredibly sensitive. What happens if they inadvertently hit a rock by mistake when probing in the soil? Is this the human equivalent of banging your funny bone?

Several different types of touch receptors are sensitive to pressure, movement, vibration, texture and pain. They differ in their appearance (under the microscope) and their distribution on the bird's body. Just as in humans, which have many more touch receptors on the fingertips than on the back of the hand, birds, which have touch receptors all over their body, have more in their beak and on their feet. Allopreening is regulated by Herbst corpuscles alone, but the manipulation of food in the bill is regulated by several different types of touch receptors and free nerve endings all working in concert.[14]

---

* American neurosurgeon Wilder Penfield (1891–1976) devised a human figure – a homunculus – whose features reflect the amount of brain tissue devoted to sensory functions in the rest of the body. The hands, lips and tongue are especially sensitive and therefore feature large on Penfield's homunculus.

Highly social bird species that breed either close together in colonies, or co-operatively like babblers and woodhoopoes, spend a lot of time allopreening. Why? A simple explanation for species like the zebra finch is that allopreening is a way of maintaining a pair bond. Just watching a pair of zebra finches nibbling each other's napes, they look as if they are in love. Indeed, this is the very reason why the small parrots called lovebirds are so named. In the past there was a tendency to assume that almost any behaviour that occurred between partners – preening, billing and mutual feeding – served to 'maintain the pair bond', but I have always found this an incomplete explanation and until very recently there was little hard evidence that behaviours like these help to maintain pair bonds.

Another explanation for allopreening in birds – and allogrooming in primates – is that it serves a hygienic function, removing dirt or parasites. The evolutionary logic is straightforward: it would pay you, for example, to remove a tick from your partner, if only because it would reduce the chances of you being infested yourself. Removing a tick from your partner may also reduce the chances of it damaging your mutual offspring. In birds at least, there are two reasons for thinking allopreening has a hygienic function. First, the behaviour is usually directed towards those parts of a bird's plumage that it cannot preen itself: the head and neck. Second, allopreening is particularly common in species that live in close proximity. The record-holder for high-density living is the common guillemot, which in flat terrain breeds at densities of up to seventy pairs per square metre, and in close bodily contact with its neighbours – the ideal situation for external parasites like ticks to creep from bird to bird. Guillemots also engage in a lot of allopreening, both with their partner, and also with their immediate neighbours with whom they are in direct bodily contact.

On Skomer Island I have hardly ever found a tick on the

hundreds of adult guillemots I have handled, and I only occasionally find them on the breeding ledges. However, at Funk Island, which I visited in 1980, there are around half a million pairs of guillemots and the gravel on which the birds breed was literally heaving with ticks. Sadly, I didn't have the opportunity to see how badly the birds were infested or whether allopreening was instrumental in removing ticks. However, one anecdote in particular suggests that allopreening may be important. Not long after the *Torrey Canyon* supertanker disaster in 1967 – in which many thousands of seabirds, including guillemots, died from being caught up in the ensuing oil slick – small numbers of survivors were kept in captivity in an effort to find ways of cleaning their plumage. One of the researchers involved in that study told me that he noticed a guillemot with a tick infestation – ticks embedded in the skin on the back of the bird's head – and how the other birds in the group fell over themselves to preen the infested individual. Clearly, the sight of a tick on the plumage was a powerful stimulus. In another study, Mike Brooke of Cambridge University showed that allopreening greatly reduced the number of ticks on wild macaroni and rockhopper penguins.[15]

Primates and socially living birds have much in common. In primates, any kind of stressful interaction, such as an attack by a more dominant individual, is often followed immediately by the victim seeking grooming, as though for reassurance. Humans do the same: we might touch someone lightly on the arm or shoulder in a gesture of reassurance or comfort. Among European magpies that I studied around Sheffield, allopreening was sufficiently rare that I made a note whenever I did see it. As in many other birds, it only ever occurred between pair members, but more interestingly it happened only after another magpie had made an aggressive incursion into their territory. Typically the intrusion resulted in a territorial skirmish, after which the pair would retreat to a tall tree, sit close together and the female then allopreened her partner – very rarely the other way round. The association with a stressful social encounter was therefore very obvious, and it is even more obvious in another bird, the African green woodhoopoe studied by Andrew Radford and Morne du Plessis.

With its spectacular iridescent green-purple plumage and scarlet down-curved bill, the green woodhoopoe is a highly social, co-operatively breeding bird. It lives in groups of six or eight individuals, comprising a breeding pair and several helpers, which are usually young from previous breeding seasons. Each night the entire group roosts together in a tree cavity, rendering them vulnerable to picking up ectoparasites from one another, so allopreening may serve a hygienic function. This seems particularly plausible since, as in other birds, the preening focuses on the head and neck. In addition, though, allopreening has a clear social function. Conflicts with neighbouring woodhoopoe groups are common and are invariably followed by allopreening between group members, just as with magpies. Allopreening under these circumstances, however, is focused on the body plumage rather than the head. The more intense the woodhoopoes' battle with their neighbours, the more intense the allopreening afterwards. Losers in intergroup conflicts also allopreened more than individuals in the winning group, presumably because losing was more stressful than winning. These birds allopreen a lot, up to 3 per cent of their day, and, as in primates, preening or grooming seems to reinforce particular social relationships.[16]

The only study so far that has explored the link between allopreening and stress reduction in any bird species has been conducted on ravens, and seems to confirm what has been found in primates: ravens that allopreened each other more often produced less of the stress hormone corticosterone. More studies are needed before we can be confident that this is a general phenomenon in birds, but my guess is that it is.[17]

The way allopreening occurs in guillemots, magpies, ravens and woodhoopoes obviously involves touch receptors in the recipient's skin. As with ourselves, a bird's skin has lots of different receptors that are sensitive to pressure, pain, movement and so on, but birds also have specially modified feathers that probably play a central role in allopreening.

There are three types of feather. The most abundant and obvious are the contour feathers: these include the long, strong wing and tail feathers, but also the short feathers that cover the body and rictal bristles around the mouth. The second type are fluffy, down feathers, lying out of sight under the contour feathers close to the body. Their role is to act primarily as insulation, hence their effectiveness in a down-filled sleeping bag or jacket. The third type of feather is much less familiar and you are likely only to have noticed them if you have ever plucked a bird like a chicken or a pigeon. Once all the contour and down feathers have been removed, what's left are the filoplumes, fine hair-like feathers sparsely dotted over the entire body surface and always rooted close to the base of a contour feather.

Filoplumes consist of a shaft, sometimes with a tiny tuft of barbs at the tip, and, like the down feathers, they are usually hidden beneath the contour feathers. In some songbird species, though, the filoplumes protrude beyond the contour feathers, as on the nape of the chaffinch or on the back of the eponymous hairy-backed bulbul. In others, the filoplumes have been co-opted as display structures, notably in cormorants, where they form the crest, but most spectacularly in the whiskered auklet. This small North Pacific seabird – it weighs just about 120 g – is extraordinarily beautiful during the breeding season, with sooty black plumage offset by a stunning white iris with a pinprick pupil, and a cluster of facial ornaments, a black, forward-pointing crest made up of modified contour feathers, and three tracts of silvery filoplumes. One set of filoplumes runs from in front of the eye down the neck, the second originates behind the eye and also runs down the neck, parallel to the first, and the third set lies above the eye and projects like antennae a few centimetres behind the head. The birds are nocturnal at the colony and, as in other auklet species, their facial ornaments probably play a role in mutual mate choice. But they also operate like the whiskers of a cat, helping the auklets avoid collisions when they disappear underground into the total darkness of their rocky breeding crevices.[18] They may do even more than this, for the whiskers (technically, vibrissae) of rats and other mammals are so sensitive that they can distinguish between smooth and rough textures, as well as objects of different sizes.[19]

For a long time the function of regular filoplumes was unknown. Indeed, a major dictionary of ornithology published in 1964 referred to them as 'degenerate, functionless structures',[20] this despite the fact that in the 1950s a German researcher, Kuni von Pfeffer, presciently proposed that filoplumes transmit vibrations via touch sensors, allowing birds to monitor and adjust their feather postures. She was right: the filoplumes are highly sensitive and when moved trigger a nerve impulse alerting the bird so that it can readjust its plumage.[21] Filoplumes must play a particularly important – albeit indirect – role in social displays. Just think of the staggering variety of feather postures birds use, including the fan-like opening of a peacock's train, the snapping of a manakin's wing feathers, the flamboyant fluffing of a displaying great bustard male and the sleeked plumage of an intimidated blue tit. The sensitivity of the filoplumes means that must also be important during allopreening, either by being moved by the allopreener directly, or indirectly, by the allopreener touching nearby contour feathers.

Before we leave the filoplumes, I should mention some similar, but more obvious, structures. First, in a number of birds, most obviously nightjars, oilbirds and flycatchers, on the corners of the mouth is an array of stiff, hair-like bristles. These are modified contour feathers, called rictal (mouth) bristles, and the presence of a well-developed nerve supply at their base betrays their sensory function. In nightjars and flycatchers the bristles help them catch flying insects. In the case of oilbirds, which are nocturnal, the bristles help them in flight to pluck fruit from forest trees in the dark. Second, certain frogmouths and potoos (nocturnal, nightjar-like birds of the tropics), kiwis and some seabirds, like the whiskered auklet, have crests or long wispy feathers on the tops of their heads. These are probably modified contour feathers rather than filoplumes, but like rictal bristles and filoplumes they probably also serve a sensory function. A recent study has confirmed this by demonstrating that birds with facial plumes are much more likely to live in complex habitats such as dense vegetation or tunnels or burrows rather than in the open, suggesting that the plumes function much like the whiskers of rats and cats and help them avoid bumping into obstacles.[22]

When Goujon discovered the bill-tip organ in parrots in the nineteenth century he said that he had also seen similar structures on the bills of wading birds such as snipe and sandpipers, species that probe in sand or mud for food. As a boy I was an avid collector of bird skulls and one of my most prized possessions was the skull of a woodcock, a probing bird, with enormous eye sockets and a distinctively pitted bill tip. These pits can be seen only after the leathery outer covering of the bill – the ramphotheca – has been removed.

Using their sensitive bill tips, probing birds like sandpipers, woodcock and snipe detect prey such as worms or molluscs either by touching them directly, by detecting their vibrations, or, more remarkably, by detecting changes in pressure in the mud or sand.[23]

Ingenious experiments by the Dutch ornithologist Theunis Piersma and his colleagues in the 1990s showed how red knots were able to detect tiny immobile bivalves (like mussels and clams) hidden in sand. When the bird pushes its beak into wet sand it generates a pressure wave in the minute amounts of water lying between the sand grains. This pressure wave is disrupted by solid objects, such as bivalves, which block the flow of water, thereby creating a 'pressure disturbance' detectable by the bird. Rapid and repeated probing, so typical of these wading birds, is thought to allow them to build up a composite three-dimensional image of food items hidden in the sand.[24]

Piersma's red knot discoveries resonated with two New Zealand researchers, Susan Cunningham and her PhD supervisor Isabel Castro, who wondered whether something similar might occur in the bill of the kiwi, the ultimate probing bird. Just as the sandpiper's, the kiwi's bill tip is a honeycomb of pits on both the upper and lower bill, both inside and outside the mouth. Interestingly, despite his careful dissections of kiwis in the 1830s, Richard Owen seems to have overlooked these pits as he makes no mention of them, nor do they feature in the exquisite drawings of kiwi skeletons in his papers. It was Jeffrey Parker, professor of biology at the University of Dunedin in New Zealand, who first reported the unusual cluster of pits in the kiwi bill tip in 1891, describing them as 'abundantly supplied by branches of the dorsal ramus of the orbitonasal nerve'. In other words the pits are richly supplied with nerves.[25] In his *Birds of New Zealand* (1873) Walter Buller provided a beautiful description of the way kiwis forage: 'While hunting for its food the bird makes a continual sniffing sound through the nostrils, which are placed at the extremity of the upper mandible. Whether it is guided as much by touch as by smell I cannot safely say; but it appears to me that both senses are used in the action . . . That the sense of touch is highly developed seems quite certain, because the bird, although it may not be audibly sniffing, will always first touch an object with the point of its bill . . . and when shut up in a cage . . . may be heard, all through the night, tapping softly at the walls.'[26]

The orientation of the sensory pits in the kiwi's bill tip provides a further clue to the way they are used to detect prey. The bill tips of knots contain neatly stacked Herbst corpuscles in forward-facing sensory pits, an arrangement that seems to be necessary to detect pressure disturbance patterns. Other sandpiper species,[27] however, that detect prey by vibration, have outward-facing pits. Kiwis, however, have pits that face forward, outward and backward, indicating that they might use both pressure *and* vibrational cues to detect their prey. Despite the similarity in their bill structures, kiwis and wading birds are hardly close relatives, but constitute a nice example of convergent evolution in which similar adaptations evolve in response to similar selection pressures – that is, in response to the need to find food concealed beneath the surface.

There is one other 'probing' lifestyle where we might expect to find a well-developed sense of touch (and taste) – on the tips of the long tongues of woodpeckers, wrynecks and piculets.

Leonardo da Vinci was one of the first to comment on the extraordinary tongue of the woodpecker,[28] but the best early illustrations are those of the Dutch naturalist Volcher Coiter (1534–76), who also recognised that the wryneck had a similar elongated tongue.[29] Sir Thomas Browne, writing in the mid-1600s, commented on the 'large nerves which tend unto [lead into] the tongue' of woodpeckers,[30] and his ornithological colleagues Francis Willughby and John Ray, after examining a green woodpecker, said: 'The tongue when stretched out is of a very great length, ending in a sharp, bony substance . . . wherewith, as with a dart, it strikes insects.' After what was clearly a very sophisticated dissection they wrote:

This tongue the bird can dart out . . . some three of four inches, and draw up again, by the help of two small round cartilages, fastened into the forementioned bony tip, and running along the length of the tongue. These cartilages from the root of the tongue take a circuit beyond the ears, and being reflected backwards to the crown of the head, make a large bow. Below the ligament they run down the sagittal suture . . . pass just above the orbit of the right eye, and along the right side of the bill into a hole excavated there, whence they have their rise or original [origin].

They go on to to describe the manner in which the tongue is protruded and retracted and finish by saying: 'But we leave these things to be more curiously weighed and examined by others.'[31]

Just over a century later, the Comte de Buffon wrote that the bony tip of the green woodpecker's tongue 'is covered with a scaly horn beset with small hooks bent back, and that it may be capable both to hold and pierce its prey, it is naturally moistened with a viscous fluid that distils from two excretory ducts . . .'.[32]

The idea that woodpeckers impale prey on their tongue persisted and was reinforced in the 1950s by the pioneering wildlife film-maker Heinz Sielmann, who wrote that the great spotted woodpecker's 'harpoon-like tongue is particularly suited to . . . spearing insect larvae and pupae'. Re-analysis of Sielmann's footage, however, showed that larvae are *not* pierced, but simply adhere to the sticky saliva at the end of the tongue. Exactly the same behaviour was seen in a study of a Guadeloupe woodpecker from the Lesser Antilles kept in captivity for a couple of weeks. On extending its long tongue into a cavity, the bird could tell immediately – using either touch or taste – when it had made contact with a prey item, and detailed anatomical studies confirm that the tongue tip is rich in touch sensors (we don't know about taste buds, but I bet they are there). In turn, the insect larva was hardly passive on sensing the woodpecker's tongue, and either retreated or grasped on to the sides of its hole with its legs, making it difficult for the woodpecker to dislodge it. Through a combination of sticky saliva, a barbed surface and an extraordinarily prehensile tip – but no piercing – the Guadeloupe woodpecker was able to extract its reluctant prey.[33]

I am in the swamps in a little-known part of northern Florida, on the Choctawhatchee River. This is red-neck country – similar to that of the 1970s film *Deliverance*. Resting quietly in my kayak, I watch spellbound as four pileated woodpeckers chase each other noisily through the trees. The late afternoon light filtering through the olive-green leaves of the buttress-rooted cypresses is perfect and the birds seem to be enjoying themselves. They flip heavily from tree to tree, hammering and calling but offering only the occasional tantalising flash of their beautiful red, black and white plumage. I have never had such wonderful close encounters with this species before, but this isn't what I am looking for. Instead, I am with a small group of ornithologists hoping to glimpse the pileated's enormous cousin, the ivory-billed woodpecker.

Thought to have been extirpated in the second half of the twentieth century, a controversial sighting on the Pearl River in southern Louisiana in 1999 suggested that at least one ivory-billed woodpecker had survived. There have been several subsequent reported sightings in remote swamps, including some on the Choctawhatchee, but so far at least no video evidence – now regarded as essential proof of the bird's existence.[34]

The ivorybill – known also as the Lord God Bird – has an enormous chisel-shaped beak. It finds its prey by searching trees in which huge beetle larvae lie hidden beneath the bark. Once a larva is located – almost certainly by the sound of its jaws chewing on its woody diet – the woodpecker hacks at and levers out a hand-size chunk of bark, exposing the larva's retreat. Imagine how much effort this would require with a hammer and chisel and you have some sense of the bird's enormous strength. As the larva wriggles away the ivorybill flashes out its extraordinary long tongue and snares it. This slick operation is one of sensory contrasts: a bill as insensitive as steel, a tongue more tactile than your fingertips.

The ivorybill's power is legendary. In 1794, Alexander Wilson, a Scottish weaver who emigrated to North America, and who later became one of the founders of American ornithology, shot an ivorybill in North Carolina. The bird was only slightly injured and Wilson decided to keep it. As he carried the bird back to town on his horse, it cried like a baby, surprising 'every one within hearing, particularly the females, who hurried to the doors and windows with looks of alarm and anxiety'. Checking into the Wilmington Hotel, Wilson left the bird in his room while he went to take care of his horse. When he returned less than an hour later he found the bed 'covered with large pieces of plaster, the lath was exposed for at least fifteen inches square, and a hole, large enough to admit a fist, opened to the weather boards; so that in less than another hour he would certainly have succeeded in making his way through'. Wilson caught the bird and 'tied a string round his leg, and fastening it to the table, again left him, this time to search for something it might eat. As I reascended the stairs, I heard him again hard at work, and on entering had the mortification to perceive that he had almost entirely ruined the mahogany table to which he was fastened, and on which he had wreaked his whole vengeance.' The bird refused all sustenance and, to Wilson's regret, died three days later.[35]

Ivorybills nest in a cavity four or five feet deep, sculpted from the live tissue of the bald cypress, among the hardest of trees. That bill, once revered as an Indian amulet, is an extraordinarily powerful tool, mounted on a seriously reinforced skull. John James Audubon dissected the head of an ivorybill and described its seven-inch [18 cm] tongue in detail, which, like that of other woodpeckers, is armed with an exquisitely sensitive tip.[36]

Audubon also provides a first account of the ivorybill's foraging technique:

> Then, having discovered an insect or larva in a chink of the bark, [it] is enabled by suddenly protruding its tongue, covered with thick mucus, and having a strong slender sharp point furnished with small reversed prickles, to seize it and draw it into the mouth. These prickles are of special use in drawing from its retreat in the wood those large larvae, often two or three inches in length; but it does not appear probable that the bristly point is ever used to transfix an object, otherwise how should the object be again set free, without tearing off the prickles, which are extremely delicate and not capable of being bent in every direction?[37]

The skin of birds and mammals alike is sensitive to both touch and temperature. This sensitivity is especially important when birds are incubating eggs or brooding chicks, not only to ensure that their eggs and chicks are suitably warmed, but also to avoid stepping on them or crushing them. The heating device is the brood patch, an area of skin from which the feathers are lost some days or weeks before incubation starts, and whose blood supply is increased.

In some birds the brood patch plays a vital role in determining how many eggs the female lays. In the 1670s the naturalist Martin Lister conducted a simple experiment on the swallows nesting near his house – with entirely unexpected results. As each egg was laid, he removed it, only to find that instead of laying a normal clutch of five eggs, the female swallow went on to lay no fewer than nineteen eggs. Why they should limit themselves to five when they could so obviously lay more was a mystery that was solved only later. Subsequent tests with other species gave similar results, including a house sparrow that laid 50 eggs (instead of 4 or 5), and a northern flicker that, instead of laying a normal clutch of 5 to 8 eggs, laid 71 eggs in 73 days! There are some species, however, like the lapwing, where removing eggs makes absolutely no difference to their final number of eggs laid. On the basis of this, ornithologists have categorised birds as either determinate (e.g. the lapwing) or indeterminate layers, although they have no idea why such a difference exists. The point, however, is that in the indeterminate layers, like the swallow, sparrow and flicker, egg laying is regulated through the brood patch. If eggs are removed as they are laid, there is no tactile stimulation of the brood patch and no message to the brain to limit egg laying. If the eggs are *not* removed, touch sensors in the brood patch detect their presence in the nest and then, via a complex hormonal process, allow only the 'right' number of ova to develop in the ovary.[38]

Once the clutch is complete it is crucial that the eggs are maintained at an appropriate temperature if the embryo within each egg is to develop normally. Successful incubation does not demand a constant temperature, but simply that it does not fall too low or get too high. Incubating birds often leave the nest to feed, during which time the eggs cool, but embryos are far more tolerant of brief periods of cooling than they are of over-heating. The eggs of most species are incubated at around 30–38°C and the incubating bird achieves this largely through its behaviour. Experiments in which eggs are artificially cooled or heated show that birds adjust their incubation posture – and in particular the contact between the brood patch and the eggs – to regulate the temperature of their eggs. This is true regardless of whether eggs are cooled – when the parent responds by transferring more heat to the eggs – or heated – when the parent responds by incubating more closely to draw off excess heat from the eggs.

On casual inspection the brood patch looks to be little more than a slightly vulgar patch of overly pink skin, but it is a remarkably sensitive and sophisticated organ. Birds regulate the temperature of their eggs by increasing or decreasing blood flow to the brood patch. What's more, the contact between the eggs and the brood patch triggers the release of the hormone prolactin from the pituitary gland beneath the brain, which in turn keeps the bird incubating. If the clutch of an incubating bird is removed, prolactin secretion plummets – tactile stimulation is crucial to this process as was demonstrated in a clever experiment in which the brood patch of incubating mallard ducks was anaesthetised. Even though the birds continued to incubate, because they were unable to feel the eggs their prolactin levels dropped, exactly as if the eggs had been removed.[39]

The only birds whose eggs are not incubated by body heat are brush turkeys, malleefowl, talegallas, maleo and scrubfowl (known collectively as 'megapodes' on account of their large feet, which they use for digging). Instead, they place their eggs in a mound of fermenting vegetation or warm volcanic soil (depending on the species), which they maintain at a temperature of around 33°C. In the mound-building species, such as the Australian brush turkey, the male cares for the mound, typically for months on end, opening it up to allow excess heat to escape, or adding more material if the mound is too cool. Darryl Jones, who has studied mound builders for years, told me: 'How they monitor mound temperatures is not yet fully understood. Most likely, both males and females possess a temperature sensor in the palate or tongue, as all species have been seen to regularly take a beak-full of substrate while working on the mound.'[40]

In birds that incubate their own eggs, the chicks must be sensitive both to each other (where there are several) and to their parents. The South American finfoot, also known as the sungrebe (a species

I searched for unsuccessfully in Ecuador) provides an extraordinary example of the need for parents and chicks to be aware of each other through the sense of touch. This secretive, little-known bird nests in dense vegetation along slow-moving rivers and hatches its clutch of two or three eggs after just ten days of incubation. Blind, naked and pretty helpless at hatching, the chicks are more like those of a passerine than a non-passerine bird. Remarkably, the male sungrebe carries his two chicks around in a special pouch of skin under each wing. He can even fly with the chicks. The Mexican ornithologist Miguel Alvarez del Toro, who discovered this, described how, on flushing a male from a nest he had been watching, he saw the male fly with 'two tiny heads sticking out from the plumage of the sides under the wings'. Surprisingly, the female does not have the pouches, nor do the males of the other two closely related finfoot species, whose chicks are much more developed at hatching. The male sungrebe's pouches constitute the most extraordinary adaptation, and they beg the question of what touch receptors are employed by the newly hatched chicks to ensure that they are in the right place, and by the adult male sungrebe to know that the chicks are absolutely secure before he takes flight.[41]

In the case of some brood parasitic birds, the tactile sensitivity of recently hatched chicks has a more sinister aspect. The greater honeyguide is a tropical brood parasite whose nestling disposes of its nest mates in a particular grisly manner. On hatching, the honeyguide has its eyes closed, but is armed with a needle-like structure on its downward pointing bill tip. It is this that it uses to kill the host young, allowing it to acquire all the food its foster parents bring back to the nest. Seeing this evil-looking device for the first time, I assumed that the honeyguide chick would simply pierce the skull or body of the host chicks, but this is not what happens. Using infra-red video cameras located inside little bee-eater host nests, Claire Spottiswoode watched as the honeyguide chick used its sharp beak to grasp a young bee-eater, and, like a pitbull terrier, simply shake it to death. If the host chick is tough, it can take several sessions, between which the honeyguide chick pauses to catch its breath before starting again. Because its eyes are not yet open, and it is dark inside the bee-eater nest cavity, the honeyguide chick presumably uses both movement (touch) and temperature to tell it whether more shaking is necessary. Once the host chick is dead the honeyguide ceases to respond to it, and the unfortunate parents remove it from the nest.[42]

It is well known that the European cuckoo chick eliminates any competition by pushing host eggs or chicks out of the nest directly. Like the honeyguide chick, it hatches with its eyes closed and relies on an acute sense of touch to detect and eject host eggs or young. Prior to Edward Jenner's direct observation of the nestling cuckoo's ejection behaviour, in 1788, many people thought that the adult cuckoo was responsible for the disappearance of host eggs or young. What's more, many people found it almost unbelievable that a newly hatched cuckoo chick could – or would – behave in such an apparently evil manner. However, once Jenner had alerted them, the sceptics soon witnessed the behaviour for themselves. 'A monstrous outrage on maternal affection,' Gilbert White called it in *The Natural History of Selborne*. Starting a few hours after hatching, the cuckoo chick begins to manoeuvre itself so that the host eggs or young are lodged, one at a time, in a small pit in the middle of its back, between its scapulars. Bracing itself against the sides of the nest with its legs, the young cuckoo heaves each victim up and over the side of the nest. Although it has not been investigated, the pit on the young cuckoo's back must be loaded with touch receptors, triggering the ejection response each time something the size of an egg or nestling touches it. After a few days, the cuckoo chick's ejection response wanes, by which time it has usually removed any host eggs or young, or sometimes even another cuckoo egg or chick.[43]

The main focus of my own research is promiscuity in birds: the behaviour, anatomy and evolutionary significance of avian infidelity. Since some birds copulate for long periods of time or copulate many times each day, a question I am often asked is: do birds enjoy sex?

In some species, like the European dunnock, copulation is so rapid – it has been timed using high-speed photography at one-tenth of a second – it is hard to imagine it generating much pleasure. On the other hand, so much of a bird's life is speeded up that perhaps to a dunnock one-tenth of a second is equivalent to several minutes for a human. In fact, most small birds copulate for only a second or two and show no sign of any physical pleasure in what has been euphemistically referred to as a 'cloacal kiss'.[44]

There are other birds that copulate for much longer, and still show no signs of pleasure, let alone ecstasy. The greater vasa parrot of Madagascar, for instance, has one of the most protracted copulations of any birds, up to one and a half hours, with the added complexity of a copulatory tie, exactly like that seen in dogs. On witnessing a canine copulatory tie for the first time, dog owners are often confused about what is going on, not least because the two animals face away from each other – because the male turns round. In the vasa parrot the copulatory tie is rather more polite in that the two animals remain perched side by side, the male nibbling his partner's head feathers (and appearing to whisper sweet nothings to her), while locked in their genital embrace. Strictly speaking, the male vasa parrot (unlike a dog) has no penis, but he does have a large, globular cloacal protrusion that, once inside the female, becomes engorged with blood (much as does the dog's penis), effectively locking the male inside the female's cloaca. The two birds sit side by side, but with little movement and even less indication of pleasure. The function of this unusual behaviour and its extraordinary accompanying anatomy, is, as my PhD student Jonathan

Ekstrom demonstrated, sperm competition: vasa parrots are among the most promiscuous of birds.[45]

I became intrigued by this species when a colleague, Roger Wilkinson, who was then curator of birds at Chester Zoo, sent me some photographs of his vasa parrots before, during and after their protracted and bizarre copulation. Not long afterwards, and quite independently, I received a message from another colleague, Andrew Cockburn, who had been birdwatching in Madagascar and had seen wild vasa parrots copulating: 'knowing of your interest in avian copulation', his message started, and then went on to describe what was effectively the same behaviour as in Roger's captive parrots. I decided this would make an interesting research project for an intrepid and enterprising student. Jonathan fitted the bill, and it was indeed a tough project. As well as coping with high temperatures and high humidity, and climbing up into the canopy and then down through the hollow trunk into the base of gigantic baobab trees, where the parrots nested, he also acted as amateur doctor to the desperately undernourished local people.

Despite all this, he obtained some remarkable results. Briefly, the birds' breeding system is unlike any other. Females sing to attract males; males appear from the forest and copulate with females – several males over several days. The female incubates the eggs alone, but when the chicks hatch she emerges to sing again, and males appear once more, this time to give her regurgitated fruit which she takes back to the chicks. DNA fingerprinting analyses revealed that almost every chick in a brood has a different father. The remarkable thing is that the males that comprise the fathers of a brood are the fathers of offspring in nests dotted across the Madagascan forest – there are no bonds of exclusivity. Like Willughby and Ray, I am happy to leave it to others to figure out why such an unusual system has evolved. What we can be reasonably confident about is that the protracted copulation in this species has almost certainly evolved in response to the intense sperm competition mediated by the females' promiscuity. By copulating for a long time – a process facilitated by the unique copulatory tie – a male probably maximises his chances of fertilising the female's eggs. Whether the females or her several partners derive any pleasure from their copulation is not known, but to be able to perform the behaviour at all must require at least some tactile sensitivity.[46]

There is one bird, however, where sexual pleasure is strikingly apparent: the red-billed buffalo weaver, a starling-sized African bird. In February 1868, when he was preparing his book on sexual selection, Charles Darwin wrote to his favourite cage-bird informant, John Jenner Weir, to ask if he could 'call to mind any facts bearing . . . on the selection by a female of any particular male – or conversely of a particular female by a male – or on the allurement of the females by the males – or any such facts'. Weir wrote back immediately describing the courtship and mating behaviour of several birds he had in captivity, including the buffalo weaver, saying that there was 'nothing particularly striking' about this species.[47]

How wrong he was. For lying just in front of the male buffalo weaver's cloaca is a false penis: a two-centimetre-long fleshy finger-like appendage. Seeing the bird going about its everyday business you wouldn't know there was anything unusual about its anatomy, for the false penis is obscured by the black body feathers. Holding the bird upside down in your hand, and blowing gently on its underside, reveals the full glory of this bizarre structure. First described in the 1830s by prolific French naval apothecary and naturalist René Primevère Lesson (1794–1849), buffalo weavers are unique among birds.

Fascinated by Lesson's account, and one from the 1920s by the Russian ornithologist Petr Sushkin, I decided to investigate further, convinced that this extraordinary structure must have evolved in relation to sperm competition. The first step was to see one for myself and by a fortunate coincidence I learned that the museum in Windhoek, Namibia, had a specimen I could have. The pickled specimen that duly arrived in the post was perfect: a male in full reproductive condition. The accompanying note explained that these were 'trash' birds in Namibia, considered a nuisance by farmers for building their enormous stick nests on windmills and disrupting the essential business of pumping water out of the ground on to the dry desert soil. My dissection confirmed everything that Sushkin said: the false penis was a rigid rod of connective tissue, with no ducts or tubes, no obvious blood supply, and, according to previous accounts, no nervous tissue either. This was odd, for the outward appearance of the phalloid organ screamed tactile sensitivity. Rarely in all my research on the reproductive biology of birds had I encountered a more pointed symbol of male virility.[48]

Significantly, my dissection revealed that the bird had relatively large testes, a sure sign of female promiscuity and rampant sperm competition. One thing led to another and, before I knew it, I had a project on buffalo weavers in Namibia, with an enthusiastic young research student, Mark Winterbottom, conducting the fieldwork. At first glance, buffalo weavers seemed very easy to study. The birds were common and in some areas almost every windmill supported their conspicuous thorny nests, which were wonderfully accessible. Less conveniently, they also nested in acacia trees, including the one overhanging the house we rented on a game farm. Waking at first light each morning to the calls of male buffalo weavers was wonderful and seemed almost too good to be true.

It was. Nests – sometimes a metre or so across and containing multiple chambers – were often owned by two males operating as a team. This is unusual for birds of any kind and an arrangement that seemed to predispose them to sperm competition. The several nests above our house were occupied by several teams, or coalitions, as we called them, of males which we caught and colour-ringed so that we could see who was who. But there were no females. Males spent their early morning at the nest, adding a few sticks and performing the occasional display or squabbling with one of the others. Then, one morning, with no warning, our males suddenly launched themselves into a frenzy of display, fluttering, bowing and calling, as a small party of females flew overhead. They didn't stop and our males' enthusiasm evaporated as rapidly as the females disappeared. Eventually it dawned on Mark and I that the buffalo weavers' breeding system was an opportunistic one, completely dependent on females taking a fancy to a group of males (or their nests) and deciding to stay and breed. The birds above our house were obviously hopelessly unattractive for, during that first four-month-long field season, no breeding occurred.

Elsewhere on the farm things were better and at another colony we soon witnessed the arrival of a group of females and the extraordinarily rapid onset of breeding. But it was copulation we were primarily interested in: how exactly did the male deploy his false penis?

The local black farm workers told us we were wasting our time for they knew why the males possessed this structure: it was a device, they said, by which the male carried thorny acacia twigs during nest construction. Our extensive observations, however, provided no evidence whatsoever for this. The local people must have known this, too, so it is curious that that particular bit of avian folklore persisted.

Witnessing copulations was tough. One morning I saw a female leave her nest chamber in what I recognised to be a purposeful way. Flying fast and low over the ground away from the colony, her unusual flight instantly alerted not only me, but also one of the two male nest owners, who immediately followed her. The two birds flew about 200 metres and landed side by side on a low branch of an acacia. I followed, too, but in 40°C heat running was hard work. Sweating profusely and barely able to hold my binoculars steady, I watched the two birds bobbing up and down beside each other in a kind of mutual display. Initially they bobbed out of synchrony, but soon they were bobbing in unison, faster and faster, and building up to a climax. Just at the moment when I thought the male would mount the female, and I would get to see what happened with the false penis, the female took flight again. The male followed; I followed, too, and we all repeated the same performance, but no cigar. They flew off again and again and I eventually lost them. They were indifferent to my presence, so it wasn't me scaring them off: this was simply the female's elaborate way of testing the male.

Over the three years of study Mark and I witnessed only a handful of copulations. Most were preceded by the synchronous bobbing display, and all the copulations were extraordinarily protracted. The male clung to the female's back, leaning backwards in an extremely unusual posture, flapping his wings to maintain balance while maintaining what looked like vigorous cloacal contact. The female, on the other hand, seemed almost to be in a trance, stoically enduring the male's endless bump and grind. What was most frustrating, though, was that we simply could not see what was going on with the false penis – we were too far away, and there were too many feathers in the way. If we were to crack this, watching wild weavers wasn't the answer. We needed to be able to observe them in captivity.

As a boy I was an avid birdkeeper and I remember looking at the ads in *Cage & Aviary Birds*, the UK birdkeepers' newspaper, and seeing buffalo weavers offered for sale. But times change and when I looked again, some thirty years later, there were no buffalo weavers on offer. Undeterred, we decided to catch some of the Namibian birds and take them back to Britain. I cannot quite believe we did this now: applying for permits, arranging air transport, veterinarians' certificates of health and so on, and I suspect that it was only because the birds were considered a pest species that we were allowed to take them. We actually sent the birds to southern Germany, to the Max Planck Institute for Ornithology, where some of my colleagues worked and where a technician, Karl-Heinz Siebenrock, was an enthusiastic and expert birdkeeper.

The birds – twelve males and eight females – were soon building nests out of the hawthorn branches that Karl-Heinz provided as a substitute for the spiny acacia they would normally use. I was optimistic that the birds would at least copulate. Before starting the study I had visited Chester Zoo where Roger Wilkinson (he of the vasa parrots) was curator of birds, and where they had three male buffalo weavers in a very large aviary. We had been to look at them, and Roger had even invited us to bring our captive birds there (I declined because I felt that the warmer summer climate of southern Germany might be more conducive to reproduction). As we walked into the zoo's giant aviary looking for the buffalo weavers among the (inappropriately) lush tropical vegetation, an unusual movement caught my eye and, lifting my binoculars, an extraordinary sight met my gaze. One of the buffalo weavers was copulating vigorously and repeatedly with a small, slightly dazed-looking dove. On and on the weaver went, copulating with the dove which squatted low and clung on to the branch for dear life. A lack of female buffalo weavers was obviously causing some frustration, but this casual observation suggested that males were both highly motivated to copulate, and did so in a protracted manner.

Our captive males were just as enthusiastic, but they did have the additional stimulation of genuine female buffalo weavers. Mark stayed in Germany to observe, sending me regular reports on his and the birds' progress. In fact, once the males were fired up and in breeding condition, their sexual enthusiasm knew no bounds. One of the things we wanted was semen samples, and, previously, using the much more modest zebra finch as our study species in captivity, we had developed a novel technique for obtaining some. Presenting a male zebra finch with a freeze-dried female mounted in a soliciting posture was often sufficient to encourage him to court and copulate, allowing us to collect his semen from the false cloaca with which we had fitted the female. I suggested that Mark try something similar using a dead female buffalo weaver we had found. The result, Mark informed me, was spectacular. Males mounted the dummy female immediately and went through their entire, lengthy copulation performance, and provided us with the much-needed semen samples. Later, when Mark showed me a photograph of his mounted female, I was horrified: it was a mere caricature of a bird, with a wire frame for a body, topped by head and wings. But it did the trick, and the males couldn't resist her.

The male buffalo weavers' unbridled sexual motivation was a godsend since it meant we could handle them without seriously disrupting their activities in our quest to understand the function of their phalloid organ. Almost any other species would have given up attempting to breed, but not the buffalo weaver. There were plenty of copulations with the real females, and, using a variety of techniques, Mark showed conclusively that, contrary to my expectations, the phalloid organ was not inserted into the female's cloaca during copulation. First, close-range video provided no evidence for penetration of the female; second, in dummy females fitted with a small piece of sponge inside their artificial cloaca, the sponge was never displaced during copulation; and third, the male's phalloid organ was rarely damp after copulation, whereas a model phallus gently inserted inside a female usually was.

The most surprising result of all was that, after the full thirty minutes of vigorous venery, the male buffalo weaver appeared to experience an orgasm. This was unheard of: no other bird in the world was known to climax. In a state of high excitement, Mark phoned from Germany to tell me. I was sceptical at first: 'How do you *know* the male is experiencing an orgasm?' Indeed, how would one tell whether the male of any other species experienced ecstasy in a similar way to ourselves? The way Mark discovered the answer may sound weird, perverse even, but biologists sometimes have to do funny things in their search for the truth.

Reasoning that what the male was doing was simultaneously stimulating both himself and the female by rubbing his phalloid organ around her cloacal region during his extended mounting, Mark decided to massage a male in the hand for the same time to see what happened. After twenty-five minutes of manipulation Mark gently squeezed the phalloid organ. The result was spectacular: the wingbeats slowed to a quiver, the entire body shuddered, the feet clenched tightly on to Mark's hand and the male ejaculated.[49]

Here was as convincing evidence as one was ever going to get that birds – well, the buffalo weaver at least – have a well-developed sense of touch in their genital region. The result flew in the face of those early researchers who failed to find any evidence of nervous tissue in the phalloid organ. How could there not be? To elicit such a dramatic response, there had to be some sensory mechanism inside the phalloid organ. It prompted me to have another look.

Using two male and two female buffalo weavers shot by farmers, I sent their phalloid organs to a neurobiologist, Zdenek Halata, in Germany. After making thin sections for examination under the light microscope and ultra-thin sections for examination with an electron microscope, Zdenek looked for nervous tissue. It was there: obvious in the males, but much less so in the females, comprising free nerve endings and touch-sensitive Herbst corpuscles (albeit much smaller than he had encountered on other parts of the body in other species). It was difficult to say more than that, but perhaps it was enough.

In human males, orgasm involves free nerve endings and other touch sensors, and a lot more besides. Indeed, the orgasm has been defined as 'an integration of cognitive, emotional, somatic, visceral, and neural processes' – or more poetically as a 'shower of stars'.[50] Intriguingly, in human males the sensory receptors in the penis are not essential since men who have lost their genitalia through warfare or accident are sometimes still able to experience orgasm.

Our main question was why it was necessary for the male buffalo weaver to experience orgasm – and surely, after all that stimulation, the female should have enjoyed an orgasm, too? Perhaps she did, but there were no outward indications to suggest that she had.

Perhaps the most significant question for us was what the advantage of such protracted copulation was for the male. It seemed fairly clear that the phalloid organ has evolved in response to female promiscuity. Our molecular results showed that the two males in a coalition shared paternity, and that females also copulated outside the coalition, so sperm competition was rife. One possibility was that males used their phalloid organ to persuade females to retain their sperm, and the greater the degree of physical stimulation, the more likely they were to do so. In other words, males were trapped in a kind of arms race, to see who could stimulate the females the most through a combination of protracted courtship, a special organ and prolonged mounting. We had no way of testing this with buffalo weavers, but studies of a promiscuous beetle showed that exactly this kind of phenomenon was possible. After inseminating the female, the male beetle performed a kind of copulatory courtship by stroking her with his legs before dismounting. When the researchers prevented males from performing copulatory courtship, the female beetle retained significantly less of that male's sperm.[51]

To sum up, it is clear that the sense of touch in birds is better developed than we might imagine, but I cannot help feeling that researchers have barely scratched the surface, so to speak. There is clearly much more to discover. Sadly, Billie is long gone, but if the opportunity to have another tame zebra finch or other bird arises, I would jump at the chance since it would be relatively simple to devise some simple, non-invasive tests to further explore their tactile world.

4

Taste

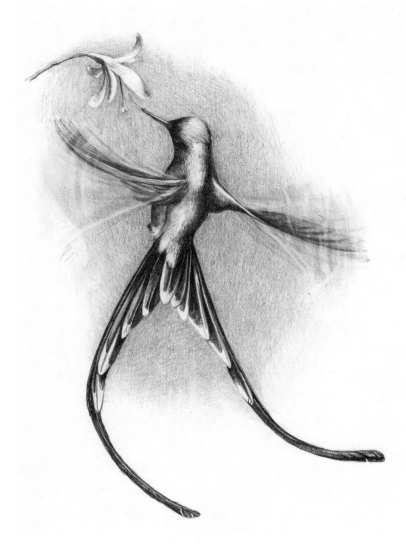

Hummingbirds can taste the sugar concentration in nectar. Here a
long-tailed sylph licks nectar from a flower – note the tongue protruding
from the bill tip: the taste buds are inside the mouth not on the tongue.

These facts and many more of a similar kind . . . fully authorize us, we think, to conclude, that some birds at least are endowed with the faculty of taste; though this is expressly or partially denied by certain authors distinguished for accuracy of observation.

James Rennie, 1835, *The Faculties of Birds*, Charles Knight

One morning in 1868, John Weir, an enthusiastic amateur bird-keeper, goes into his aviaries with some caterpillars for his birds to eat. They love natural food like this and prefer it to their usual artificial diet, but on this occasion Weir notices that, while the birds rapidly consume some caterpillars, they leave others untouched. Looking more carefully, he realises that the caterpillars the birds are eating are all cryptically coloured while those they are avoiding are brightly coloured. Wondering whether this has anything to do with their taste, Weir later offers his birds some caterpillars of the ermine moth, which he knows to be distasteful. Most of the birds refuse even to try them, but the one or two that do try them reject them immediately, shaking their heads and wiping their beaks, and are clearly agitated. Weir has witnessed the first evidence that birds have a sense of taste.

John Weir conducted this experiment at the request of Alfred Russel Wallace, with Charles Darwin the co-discoverer of natural selection. Darwin and Wallace were both fascinated by the colour of animals – and by birds in particular – and by the fact that males are usually brighter than females. Darwin's explanation for this difference between the sexes lay in what he called sexual selection – that females preferred to mate with brighter, more attractive males.[1]

The colours of one group of animals, however, the caterpillars of butterflies and moths, could not be explained in this way because as

larvae they are sexually immature and incapable of reproduction.
Searching for an alternative explanation – he was preparing his
book on sexual selection at the time – Darwin sought the advice of
Henry Bates. A superb naturalist, Bates had travelled extensively in
the Amazon in the 1850s and written detailed accounts of the insects
there. Bates in turn suggested that Darwin ask Alfred Wallace, who
had been with him in South America.

Wallace wrote to Darwin on 24 February 1867 saying: 'I saw
Bates a few days ago and he mentioned to me this difficulty of the
catterpillars [sic]. I think it is one that can only be solved by special
observation.' Wallace then speculates that:

> Birds . . . I presume are great destroyers of catterpillars. Now
> supposing that others, not hairy, are protected by a disagree-
> able taste or odour, it would be a positive advantage to them
> never to be mistaken for any of the palatable caterpillars,
> because a slight wound such as would be caused by a peck of
> a bird's bill almost always I believe kills a growing catterpillar.
> Any gaudy and conspicuous colour therefore, that would
> plainly distinguish them from the brown and green eatable
> catterpillars, would enable birds to recognise them easily as a
> kind not fit for food, and thus would escape seizure which is
> as bad as being eaten.[2]

He continues: 'Now this can be tested by experiment, by any one
who keeps a variety of insectivorous birds. They ought as a rule to
refuse to eat and generally refuse to touch gaudy caterpillars, and to
devour readily all that have any protective [i.e. camouflage] tints. I
will ask Mr Jenner Weir of Blackheath about this.'

John Jenner Weir and his brother, William Harrison Weir, were
knowledgeable and trustworthy birdkeepers from whom Darwin
regularly sought information. In response to Wallace's request,

John Weir, who was an accountant by profession, conducted the necessary experiments in his spare time and early in 1868 he reported to Darwin what he had observed.

John Weir's observations were verified by the eminent entomologist Henry Stainton, who told Alfred Wallace in 1867 how, after 'mothing' [i.e. moth trapping], he was accustomed to throwing all the common species to his poultry. On one occasion a brood of young turkeys greedily consumed the moths he threw to them, but 'among them was one common white moth. One of the young turkeys took this in his beak, shook his head and threw it down again, another ran to seize it and did the same, and so on, the whole brood in succession rejected it.' The white moth, the ermine, was the adult of the larvae Weir's birds had found so unpalatable.

The pioneering efforts of Wallace, Weir and Stainton have since been amply confirmed by more recent researchers, including Christer Wiklund, a behavioural ecologist (and Bob Dylan aficionado) from the University of Stockholm in Sweden. In the 1980s Wiklund and colleagues used:

> naive individuals of four different bird species, great tit, blue tit, starling and common quail, and showed that it seems a general phenomenon that birds do have a sense of taste and release distasteful and aposematic [warningly coloured] insects more or less immediately after taking them in their beak (and without harming the insect – presumably not out of sentimentality (or generosity) but rather so as not to get all that nasty stuff in their mouth).[3]

All these observations provide clear evidence that, using their sense of taste and vision, birds associate the appearance of their prey with its palatability. Warning colouration (also known as 'aposematic' colouration) has since been found in a wide range of animals, including insects, fishes and amphibia, and, as we'll see, in birds as well.

Even by Darwin's day, the issue of whether birds possess a sense of taste had long been debated. On the one hand, the hard beak of birds is so different from our own soft, sensuous mouth that it is difficult to imagine birds of any kind being able to taste much. The human mouth is a remarkable structure; soft and moist, with a large, fleshy tongue acutely sensitive to gustatory, thermal and tactile sensations, both while eating and during erotic kissing. The difference between our own mouth and that of birds could hardly be greater: the bird's beak is hard, often sharply pointed, and the inside of the mouth itself looks distinctly unsensuous. In most species the tongue is a hard, understated, arrow-like structure, lying within the lower jaw, and at a first sight appears to offer minimal purchase for many taste buds. Also, because birds lack teeth and do not chew their food but swallow it directly, the impression is that they have no sense of taste. Add to all that the fact that a bird's beak limits any facial expression associated with taste – pleasure or disgust – it is hardly surprising that the general impression is that birds have little or no sense of taste.[4]

Taste buds in humans were first described in the nineteenth century. Long before that, though, people had been fascinated by taste. Aristotle believed that the sensation of taste was transmitted from the tongue via the bloodstream to the heart and liver, which, in the fourth century BC, was considered both the seat of the soul and the source of all sense perception. The Roman anatomist Claudius Galen (AD 129–201) later debunked Aristotle's idea by tracing the nerves in the tongue to their origin in the base of the brain. The discovery of taste 'papillae' (nipple-like structures) on the human tongue by the Italian anatomist Lorenzo Bellini in 1665 was almost certainly inspired by Marco Malpighi's (1628–95) discovery of papillae on the tongue of an ox the previous year. Bellini's description is wonderfully enthusiastic: 'Many papillae are evident, I might say innumerable and the appearance is so elegant . . .' He describes them as appearing like 'innumerable mushrooms emerging between fine, densely standing blades of grass . . .' True taste buds – microscopic nerve endings – were not discovered for another two centuries, in frogs and fishes and then in humans, in the 1850s and 1860s. The fact that they were associated with the papillae on the tongue strongly suggested they were involved with taste.[5]

The Scottish naturalist James Rennie, writing in 1835 in his book *The Faculties of Birds*, says that while 'some birds at least are endowed with the faculty of taste' this is 'expressly or partially denied by certain authors distinguished for accuracy of observation, such as Colonel Montagu and M. Blumenbach because in several species the tongue is "horny, stiff, not supplied with nerves, and consequently unfit for an organ of taste"'. But, as Rennie perceptively points out, 'it does not follow, that because the tongue in most other animals is the chief organ of taste, that birds . . . cannot discriminate their food by taste, since other parts of the mouth may perform this office.'[6]

Rennie was almost alone at that time in imagining birds to have a sense of taste, but a moment's reflection makes it improbable that birds could function without one. Taste is essential for discriminating between edible and non-edible (or dangerous) food items. Nonetheless, sixty years later, writing in his massive *Dictionary of Birds*, Alfred Newton says:

> The tongue is commonly supposed to be the chief organ of taste; but it is certainly not so in birds . . . It is true that the tongue of birds is very rich in sensory bodies . . . which are terminal organs of the sensory nerves; but these corpuscles are

frequently embedded deeply in and beneath the impervious horny sheath, so that they cannot serve as organs of taste though they may act as organs of touch . . .[7]

Of course, taste buds were eventually found in birds – how could birds not have had a sense of taste? – and for a while the definitive overview of the field was one published in 1946 by Charles Moore and Rush Elliot. According to them, the few taste buds birds possessed were restricted to the tongue, a view that subsequent researchers accepted without question.[8]

Fast-forward to the 1970s, and the University of Leiden in the Netherlands, where Herman Berkhoudt is a young PhD student. His research topic is the microscopic structures associated with the sense of touch in the beaks of birds. One day in January 1974, while supervising two students engaged in the familiar anatomist's exercise of constructing a 3-D image from a series of thin two-dimensional sections – in this case of a duck's head – he made an exciting discovery.

Berkhoudt had enlarged the sections they were looking at by projecting them on to a table so that they could be easily traced, and, as one particular image appeared, he noticed something unusual. At the very tip of the duck's beak there were 'strange, ovoid clusters of cells leading to a pore inside the beak tip'. He told me: 'At that moment I realised that I [had] found taste buds. It gave me quite a dose of adrenalin!' This was new. All previous studies of birds' taste buds said that they occurred only on the tongue or towards the back of the mouth.

Berkhoudt's discovery diverted him from his original research topic of touch to taste. A few years earlier some colleagues in his own department had shown that mallard ducks possess a remarkable ability to discriminate between ordinary peas (which they loved to eat) and peas made to taste unpleasant, simply by grasping them with the bill tip. The ducks never got it wrong; they could always pick out the palatable peas. Figuring out precisely how they did this became the main objective of Berkhoudt's studies.

Over the next few years, his meticulous microscopic examination of the mallard's mouth revealed a total of some four hundred taste buds in the upper and lower jaws, with none – rather oddly, given previous studies – on the tongue itself. The taste buds occurred in five discrete clusters, four in the upper and one in the lower jaw. The next stage was to see why the taste buds were positioned where they were. To investigate this, Berkhoudt used an ingenious technique of high-speed X-ray filming of ducks picking up and swallowing food. This revealed that the points where the bird grasps food (the bill tip), and where food comes in contact with the inside of the mouth as it is moved towards the throat, coincides exactly with the position of the taste buds, providing a clear explanation for the birds' ability to distinguish true peas from artificially unpleasant-tasting ones.[9]

An important part of any PhD study is the need to be thoroughly familiar with what has been published previously on one's particular research topic. This is an essential part of scholarship, without which it is all too easy to end up reinventing the wheel. In addition, knowing what earlier researchers have done enables one to build on their discoveries and avoid the pitfalls identified by their work. Sometimes, though, if the earlier literature is in another language, it can remain inaccessible. Fluent in German, Herman Berkhoudt was amazed to find a series of papers from the first decade of the twentieth century that had been completely overlooked by all previous taste researchers. The first of the unknown papers was by Eugen Botezat at the University of Czernowitz in the former East Germany, who found taste buds on the tongues of young sparrows in 1904. The second was Wolfgang Bath, at the University of Berlin, who confirmed the presence of taste buds in birds in 1906 and – significantly – showed that they were not confined to the tongue, as Botezat had said.[10]

A little disappointed that his own results were not quite as novel as he first thought, Berkhoudt was nonetheless intrigued by what these German anatomists had found. He also realised that his own discoveries opened up some exciting research opportunities, and he made good use of them. Employing new, efficient ways of finding and counting taste buds, Berkhoudt plotted their distribution in the mouths of ducks. Because earlier researchers were unaware of Botezat's and Bath's papers, and persisted in focusing their efforts on the tongue, they had seriously underestimated the total number of taste buds that birds possessed.

We now know that the chicken has 300 and, from Berkhoudt's work, that the mallard has about 400; Japanese quail have just 60 and the African grey parrot has at least 300–400. But apart from these few species, we still have remarkably little information on the total number of taste buds possessed by birds. If you look at textbooks dealing with the senses, the numbers of taste buds for a variety of birds are listed, including the blue tit, bullfinch, ringed dove, European starling and an unknown species of parrot. Yet, as far as I can tell, these are all underestimations since they refer only to parts of the mouth.[11]

In most bird species the taste buds are located at the base of the tongue, in the palate and towards the back of the throat. Since saliva (or, at least, moisture) is crucial for the perception of taste, many taste buds are, not surprisingly, located near the openings of the salivary glands. Based on the limited amount of information available, birds have relatively few taste buds compared with humans (10,000), rats (1,265), the hamster (723) and a species of catfish (100,000).[12]

Despite the general assumption that there is a correlation between the amount of sensory tissue and how developed that particular sense is, the actual number of taste buds may not tell us very much about what birds can actually taste or how well they can discriminate between different tastes.

The ability of birds to distinguish different tastes was investigated in the 1920s by the scientist Bernhard Rensch and the birdkeeper Rudolf Neunzig. They screened sixty species of birds for their taste perception, simply by presenting birds with a single water container dosed with various chemicals to create the four main taste stimuli – salt, sour, bitter and sweet – that humans respond to. The birds' water consumption was compared with a 'control' group of birds provided with pure water. In later studies the experimental design was improved and the same birds were given two water containers, one in which the test substance had been dissolved, the other containing pure water. Preference for one or the other was taken as evidence that birds could taste the difference between the two containers.[13]

These studies confirmed that, despite their relatively small number of taste buds, birds respond to the same taste categories – salt, sour, bitter and sweet – as we do. (It is not known whether they respond to the most recently discovered taste category, umami – savouriness.) We also know that hummingbirds can taste differences in the amount of sugar in nectar, that fruit-eating birds can distinguish between ripe and unripe fruit – on the basis of its sugar content – and that wading birds such as sandpipers can taste the presence of worms in wet sand.[14] On the other hand, birds and humans are known to respond very differently to certain tastes. Birds seem to be indifferent to capsaicin, the substance that for us makes chilli peppers hot; indeed, in the late 1800s bird breeders fed red peppers to their canaries to turn their plumage red, and the birds ate them with no evidence of discomfort.[15] Despite this, a major article on taste in birds published in 1986 concluded: 'Research on taste in birds has been handicapped by the general assumption that they live in the human sensory world.'[16]

In 1989, Jack Dumbacher, a PhD student at the University of Chicago, made a remarkable discovery: he found the world's first distasteful bird. Jack was studying Raggiana birds of paradise in Varirata National Park, Papua New Guinea. He and his fellow students set nets to catch their birds of paradise, but, as often happens, they caught other species as well. One of the commonest bi-catch species was the hooded pitohui (phonetically: pit-oh-wheez), a bird with striking orange and black plumage. The pitohui were a nuisance, not least because they smelled and were always feisty on being removed from the net. On one occasion a bird scratched Dumbacher as he was handling it, breaking the skin. Not long after, while sucking the wound, Dumbacher became aware that his mouth had become numb. At the time he thought little of it, but when another student reported the same thing some time later he began to wonder if there was something special about the pitohui. There wasn't time that season to check, but the following year Jack took a single feather from a pitohui he had just caught and tasted it. The effect was electric. There was something extraordinarily unpleasant on the feathers.

When Bruce Beehler, Dumbacher's PhD supervisor, visited a few months later, Dumbacher told him what he had found, wondering modestly whether it might make an interesting note for a local bird journal. Beehler erupted: 'Are you telling me you've found a poisonous bird? . . . This should be on the cover of *Science*! Turn the car around! We're going back to town to get permission to study this bird!'

Bruce Beehler probably knows more about New Guinea birds than almost anyone else – he wrote the definitive *Birds of New Guinea* – and he recognised immediately that Dumbacher had made an extraordinary discovery. He was amazed that no one had previously commented on the hooded pitohui's toxic feathers – the species had been known to science since the mid-1800s, it was common locally and there were dozens of skins in museums across the world.

In fact the local people *did* know all about the hooded pitohui – they called it the *wobob*, which literally means 'the bird whose bitter skin puckers the mouth'. It was one of Dumbacher's colleagues who told him that the pitohui's unpleasant taste had previously been described in 'an old book' written by a New Zealand anthropologist, Ralph Bulmer, with a local man, Ian Seam Majnep. Old? When I checked, I discovered that the book had been published as recently as 1977. When Dumbacher checked, he was surprised to learn that, in addition to the *wobob*, local people knew of yet another distasteful New Guinea bird, this one from the highlands: the blue-capped ifrita (a species that behaves like a nuthatch), known locally as *slek-yakt*, meaning 'bitter bird'.[17]

Dumbacher wondered what the toxin on the feathers of these birds was and, by an extraordinary stroke of luck, was directed to the only person in the world who could help him find out. John Daly, a pharmacologist at the National Institute for Health, had spent years studying the toxins (so-called batrachotoxins) produced by South American dart poison frogs. As Dumbacher told me:

I was also incredibly lucky to have teamed up with the one chemist in the world who could have isolated and identified batrachotoxins easily in the lab. We were so skeptical about our initial findings (partly because . . . it seemed so unlikely that these toxins would turn up in a New Guinea bird) that we repeated the extractions on several birds before believing the results. But there they were, and after much collecting and work, we even described several new batrachotoxin compounds [from the birds] that had not previously been recovered from the frogs.[18]

The toxins in the feathers and skin of the pitohui are derived from its diet (as they are in other toxic animals), in this case from melyrid beetles. The new batrachotoxin is more toxic than strychnine. Indeed, when extracts from pitohui feathers were injected into mice, they had convulsions and died – fairly convincing evidence for toxicity.

Ongoing research by Dumbacher and colleagues revealed a total of five toxic birds in New Guinea (so far): the hooded, rusty, black and variable pitohuis and the blue-headed ifrita, all with the same toxins and all often emitting a powerful, acrid odour. The toxins may have evolved initially as a way to keep feather-eating lice at bay, and only later developed to deter larger predators. Jack Dumbacher has never seen a bird of prey attempt to catch or kill one of his distasteful birds to observe its reaction, so we don't know whether they would find it unpalatable. He has, however, conducted experiments with snakes and told me: 'Brown tree snakes and the green tree python both react strongly to the toxins and appear distressed and generally irritated by it, but we were not able to do sufficient experiments to confirm (or refute) that these snakes learn to avoid the toxins.' He also said: 'I personally suspect that the greatest benefits of the toxins accrue during nesting, and help protect the otherwise defenceless nests (eggs and young) or roosting birds from predators. An earlier description of a single hooded pitohui nest suggests that downy chicks are brightly colored, and I have always wanted to find an active nest to test for toxins, but I have never been so lucky.' Dumbacher's idea is that the substances from the adults' feathers rub off on to the eggs during incubation and help deter egg predators such as snakes.[19]

Dumbacher and Beehler duly published their paper in *Science* in

October 1992 – with a cover photograph – alerting the scientific world to the presence of distasteful, poisonous birds.[20] It prompted researchers to tell them about other birds that appeared to be toxic. These included the story of John James Audubon, who boiled up the carcasses of ten Carolina parakeets he had shot (the bird is now extinct) for his cat, specifically to see if they were poisonous. He doesn't say, but the cat disappeared, and he commented on the fact that seven cats had died the previous summer from eating 'parokeets'. The birds fed on cocklebur seeds – known to contain a toxin – so they probably were poisonous.[21]

Another intriguing example is the suitably conspicuous red warbler of Mexico, described in the Florentine Codex – the pre-Columbian account of Aztec flora and fauna – as inedible. Prompted by Dumbacher's discovery, researchers revealed that the feathers of the red warbler contain alkaloids, which, when injected into mice, caused 'unusual behaviour'.[22] This particular study is tantalisingly incomplete: a wonderful opportunity for a Mexican ornithologist and biochemist to collaborate.

Because no one has so far witnessed a predatory bird catching a pitohui or an ifrata, we simply do not know how it would respond. Would it react like Jack Dumbacher or the snakes he tested – with disgust and rejection? My guess is that it would.

New Guinea's distasteful but brightly coloured birds were similar to Darwin's and Wallace's caterpillars in which bright colours serve as a warning: *Don't eat me, I'm distasteful.* Neither Darwin nor Wallace ever imagined this might also be true of birds, mainly because so many birds – duck, woodcock, even larks and thrushes – are to us extremely tasty.

Jack Dumbacher's discovery showed convincingly that birds could be distasteful and that distastefulness was linked to bright plumage. Yet this was not without precedent, for fifty years earlier this had been a hot topic of research.

In October 1941 Cambridge zoologist Hugh Cott (1900–87) was
serving with the British Armed Forces in Egypt. He was on a week's
leave and was skinning some birds he had shot and preparing them
as museum specimens. As did so, he noticed something unusual.
Beneath the table at which he worked lay the carcasses of a palm
dove and a pied kingfisher. Hornets were feasting on the palm dove,
but ignored the pied kingfisher lying alongside it. The dove was
cryptically coloured, the kingfisher a striking black and white. This
set Cott thinking. He was already fascinated by the colour of
animals and his book *Animal Colouration*, now a classic, had been
published the previous year.[23] As Cott later said, his hornet encounter was 'a good example of the way in which a chance and quite
unexpected observation may suggest and lead to a fruitful and little
explored line of enquiry'.[24]

At this time the idea that the bright plumage of birds might serve
to protect them from potential predators was entirely novel, and
over the next twenty years Cott pursued it relentlessly. Using
hornets, cats and people as his 'tasters', together with accounts from
fellow bird-eaters, Cott assessed the palatability of species as diverse
as hoatzins, hawfinches, hoopoes and house sparrows. He concluded
that the really palatable birds, like woodcock, grouse and pigeons,
are dull- or cryptically coloured, whereas distasteful species are
more colourful – warning colouration. His discovery led to a paper
in *Nature* in 1945.[25]

Cott's study, however, is full of holes. Part of the problem, it
would be fair to say, is that the nature of scientific investigation has
changed enormously since the 1940s, and Cott's methods, which at
best seem merely quaint, are by today's standards simply inappropriate. In scoring the plumage brightness of birds, for example,
Cott used only females, ignoring the (inconvenient?) fact that males
and females are often strikingly different. He assumed (but never
checked) that males and females tasted the same. Cott also tasted

only the flesh, and cooked flesh at that, unlike Dumbacher, who (albeit accidentally) tasted the pitohui's feathers – which, after all, is what predators would encounter first. As we have seen, human senses do not necessarily provide a good measure of avian senses, so what tastes bad to us may not taste bad to a raptor or a snake. We also know that some of Cott's informants were unreliable – to say the least.[26]

It is unlikely that anyone will ever redo Cott's study using more rigorous methodology, but, as far as I am concerned, the question of association between plumage brightness and palatability across birds *in general* remains open. Since there is good evidence that plumage brightness plays an important role in avian mate choice, any reappraisal of colour and distastefulness would need to take this into account, too. On the other hand, we now know that some birds at least have a well-developed sense of taste and learn to reject certain insects on the basis of this. In principle it would not be that difficult to undertake some simple behavioural tests to discover if certain birds are distasteful to their predators. One could, for example, give some captive New Guinea raptors a piece of meat (just enough to test their reaction without putting them at risk . . .) wrapped in pitohui feathers, to see how they respond.

We can end this chapter by confirming that birds do indeed have a sense of taste. It isn't staring us in the face, so it has been under-researched, but it is there. Our knowledge of which birds have this ability is still limited and it would be wonderful if someone were to undertake a really comprehensive survey, perhaps using brain-scanning technology as a rapid way of screening a large number of species. I recognise that to some readers our lack of knowledge about what birds can or cannot taste may seem frustrating, but, as a researcher, I look on this as an opportunity. The field is wide open, with fabulous opportunities for discovery!

5

Smell

A brown kiwi. Thumbnails (*from left to right*): bill tip (*side view*) with numerous pits containing sensory nerve endings and nostril (*large opening*); cross section through the upper part of the bill showing the complex nasal region; a kiwi's brain (*beak would be to the left*), showing the huge olfactory bulb (*darker shading*).

There are certain things in the realms of ornithology which commonly pass as instinct for want of a better name, but which really are a recognizable part of a bird's economy [way of life]. All are at times rather incomprehensible, but the most perplexing is the capacity for scent – alleged by some, denied by others.

John Gurney, 1922, 'On the sense of smell possessed by birds',
The Ibis, 2, 225–53

In the mid-1500s, João dos Santos, a Portuguese missionary in East Africa (in what is now Mozambique), complained in his diary that each time he lit the beeswax candles in his tiny mission church, small birds would come in and eat the warm wax. The local people told dos Santos that the bird was 'sazu' – 'the bird that eats wax' (which he might have guessed). We now know that these were honeyguides and, writing four centuries later, Herbert Friedmann asked 'how the bird becomes aware of the presence of wax in places where apparently there are no bees . . . To this there is no satisfactory answer as yet. The possibility of their finding it by smell is very slight as birds generally have but poor olfactory acuity.'[1]

For some inexplicable reason ornithologists have found it hard to accept that birds might have a sense of smell. Ask almost any of them and they'll turn up their noses and say, no, there's not much going on in the olfactory department of a bird's brain. They are wrong, and we have to thank John James Audubon, one of the greatest bird artists ever, for setting us off on the wrong track. As a child in the late 1700s, Audubon had been told that the turkey buzzard (aka the turkey vulture) found its carrion food by means of 'an extraordinary gift of nature': an acute sense of smell. But as Audubon later observed, 'nature, although wonderfully bountiful, had not granted more to one individual than was necessary and that no one was possessed of any two of the senses in a very high state of perfection; that if it had a good scent, it needed not so much acuteness of sight'. In other words, Audubon had the strange idea that it was impossible for a species simultaneously to have two well-developed senses. When he discovered that turkey vultures were unable to smell him when he approached them, concealed behind a tree, but 'instantly flew away, much frightened' when they saw him, any idea of an acute sense of smell evaporated, and he 'assiduously engaged in a series of experiments, to prove to *myself*, at least, how far this acuteness of smell existed, or if it existed at all'.[2]

A larger-than-life character, Audubon was the dynamic, erratic and charming illegitimate son of a French sea captain and a servant girl. Born in Haiti in 1785, he moved to France at the age of six where he lived with his father and his childless wife, Anne. At eighteen his father sent him to Pennsylvania to oversee a plantation, but John James had no aptitude for farming, or, indeed, for anything that might earn him a living. Instead, he was passionate about birds, observing, shooting and drawing them. In doing so he discovered new species, made some original observations about bird behaviour and honed his artistic talents. He also found time to court Lucy Bakewell, the daughter of an English neighbour, whom he married in 1808.

Determined to make a living from his bird illustrations, Audubon took himself to the east coast of America, where, despite making many useful contacts, he failed to convince anyone of the merit of his artistic efforts. Seeking fortune further afield, Audubon set off for England in 1826, leaving Lucy and their small children behind. He was confident, cocky and proud of his skills as a field ornithologist, and his first exhibition in Liverpool was a success. No one painted birds like this – life-size, in realistic postures with every relevant feature illustrated. It was precisely because Audubon knew his birds so well that he was able to capture their essence so exactly.

Long before he set off for Britain, Audubon had tested the idea that turkey vultures had a sense of smell. His experiments consisted of hiding carcasses of various large animals and waiting to see whether vultures found them. Invariably they didn't, and Audubon concluded that unless a carcass was visible the birds could not find it. So convinced was Audubon by his results that he decided to present the details of his vulture experiments to the Edinburgh Natural History Society in 1826. The title of the subsequent paper – as long-winded as it was provocative – says it all: 'An account of the habits of the turkey buzzard (*Vultur aura*), particularly with the view of exploding the opinion generally entertained of its extraordinary power of smelling'.

The effect of Audubon's publication on the ornithological community was remarkable. It divided it, but not equally, for most of its members took Audubon's side, considering his experiments 'unanswerable' – that is, utterly convincing.[3] His disciples included William MacGillivray, Audubon's friend and ghostwriter,[4] and several other eminent ornithologists, including Henry Dresser, William Swainson, Abel Chapman, Elliot Coues and Lord Lilford. The latter two were 'sportsmen' and their evidence for 'no sense of scent' came directly from their experience as hunters. It seemed not to matter, they said, whether they approached birds upwind or not; in most cases it made no difference.[5]

Among Audubon's most enthusiastic supporters was the American Lutheran pastor and naturalist John Bachman, who repeated Audubon's experiments in the presence of 'a learned group of citizens'; they in turn then signed a document to the effect that they had witnessed the tests and were thoroughly convinced that the vulture lacked a sense of smell and was attracted to its prey 'entirely by vision'. Science by committee![6]

Of Audubon's critics the most vociferous was the astute but eccentric Charles Waterton who lived at Walton Hall, in Yorkshire.

Waterton had spent many years in South America studying natural history and was familiar with turkey buzzards, and he was utterly convinced that Audubon's experiments were flawed. Waterton was right, but his arguments were so convoluted, and his manner so strange, that the ornithological community ignored him.[7]

Audubon's experiments were indeed flawed. He made the mistake of assuming that vultures sought out putrifying, smelly carcasses and so these were what he used in his experiments. We now know that although these vultures feed on carrion, they prefer fresh carcasses and assiduously avoid those that are decomposing – hence Audubon's erroneous results. Another problem compounded the confusion. Audubon said that he conducted his experiments on what he called the turkey buzzard – that is, the turkey vulture *Cathartes aura* – but in fact it appears that the bird he was studying was the black vulture, *Coragyps atratus*, which, while similar in appearance, has a much poorer sense of smell than the turkey vulture.[8]

Further investigations designed to establish whether birds had a sense of smell reinforced the view that they didn't, even though, like Audubon's, these experiments were appallingly designed. One such investigation, performed by Alexander Hill in 1905, consisted of presenting a single domesticated turkey with two lots of food, under one of which was added some strong-smelling substance, including lavender oil, essence of anise and tincture of asafoetida.[9] The prediction was that if the turkey had a sense of smell, it would consume only the food uncontaminated by smell. Instead, the bird ate the lot. Hill's final experiment consisted of presenting the unfortunate bird with food together with a saucer of hot, diluted sulphuric acid to which he added an ounce of potassium cyanide. The resulting reaction was extremely violent and produced a cloud of deadly prussic acid that killed the turkey. From these experiments, published in the journal *Nature* no less, Hill concluded that the turkey – and by inference all other birds – had no sense of smell.

While the 'scientific' evidence seemed to preclude the possibility of a sense of smell in birds, there was plenty of anecdotal evidence to suggest exactly the opposite. In Norfolk in the late eighteenth century, the blue tit was known as the 'pickcheese' for its habit of entering dairies and eating cheese; presumably they could smell it. Hardly convincing evidence: dairies are predictable in their location, so the birds could have learned; more telling would have been if the tits visited only when cheese was being made. We don't know. In Japan some three hundred years ago, the closely related varied tit was taught to tell people's fortunes. The fortune-teller read a poem aloud after which the (tame) bird picked out a card – placed face up on a table – that matched the poem. This was a particularly difficult trick to train the bird to perform, but the owners did so by placing something burned on the back of those cards they did *not* want the bird to pick. Since it worked, this suggests that the bird used its sense of smell to distinguish between the cards. Another anecdote refers to the ability of certain waders to smell mud. The Norfolk naturalist John Henry Gurney recounts that:

> In Norfolk it is a common practice to 'fye out' a drain, that is to cleanse a 'dyke' or pasture water-course, and a very smelly operation it sometimes is. Again and again have I remarked how the attraction of the mud is sure to bring sooner or later the green sandpiper, by no means an abundant bird at any time . . . But how do they manage to discover the freshly-turned mire which is to provide them with a meal unless they smell it?[10]

More convincing are the many anecdotes about ravens sensing death. This one in particular sounds like something from a Thomas Hardy novel:[11]

In May 1871, Mr E. Baker of Merse in Wiltshire was attending
the funeral of two children who had died from diphtheria.
The road to be followed lay along the Downs for a mile or
more and the hearse had not proceeded far when two Ravens
made their appearance. These sable birds . . . accompanied
the mourners most of the way, and attracted attention by
making repeated stoops at the coffin, leaving no doubt in Mr
Baker's mind that their powers of scent had detected what was
inside them.[12]

One commentator said: 'After reading this narrative it is difficult to
treat the long-established belief about ravens as a fable; here it is
quite certain that sight could have been of no avail as the coffins
were closed, and the ravens could only have realised what their
contents were by scent.'[13]

The widespread idea that ravens foretold death accounts for their
appearance in Shakespeare's *Othello* (Act IV, Scene I): 'As doth the
Raven, o'er the infected house, Boding to all.'

The anatomical evidence was stronger still. The nineteenth
century saw huge advances in our understanding of animal anatomy.
Dissection became a passion, especially among British and German
zoologists. The most proficient in England was Richard Owen –
later Darwin's nemesis for rejecting natural selection, sticking
instead to the Establishment view that God had created all life in its
present form. Form was what it was all about, for Owen was a
superb dissector and a shameless social climber, probing and slicing
his way into the upper echelons of Victorian society.

The Victorian obsession with anatomy set the agenda for university zoology degrees for the following century and a half. As an
undergraduate in the late 1960s I dissected my way through much
of the animal kingdom: earthworms, starfish, frogs, lizards, snakes,
pigeons and rats. I loved it. The dogfish was our model organism;

week in, week out we retrieved our personally labelled dogfish from a huge vat of stinking formalin so that we could continue our dissection. The cranial nerves were especially important, emerging from the brain to control most bodily functions, but whose significance I barely understood at the time. Despite the paralysing effect of the formalin on our nostrils, the dogfish was a delightful dissection. Its skeleton, made of cartilage rather than bone, allowed us to pare away the skull – it was like slicing beans – to expose the rope-like nerves leaving the brain. The fifth nerve, the trigeminal (so-named because it has three major branches), as in all vertebrates, carries information from the nasal cavity to the brain.

It was this nerve that Richard Owen exposed in 1837 in a turkey vulture he dissected to check Audubon's assertion that the species did not find its food by smell. Owen compared the turkey vulture with a turkey, which he felt was an appropriate comparison, being the same size and 'one in which the olfactory sense may be supposed to be as low as in the vulture, on the supposition that this bird is as independent of assistance from smell in finding his food as the experiments of Audubon appear to show'. The dissection revealed the turkey vulture's trigeminal nerve to be particularly large and Owen concluded that 'the vulture has a well-developed organ of smell, but whether he finds his prey by that sense alone, or in what degree it assists, anatomy is not so well calculated to explain as experiment'. On the other hand, there were numerous anecdotes consistent with the turkey vulture having a well-developed sense of smell; one that Owen mentions came from a Mr W. Sells, a doctor in Jamaica:

> The bird is found in great abundance in the Island of Jamaica, where it is known by the name of John Crow . . . an old patient and a much valued friend who died at midnight: the family had to send for necessaries for the funeral to Spanish

Town, distant thirty miles, so that internment could not take place until noon of the second day, or thirty-six hours after his decease, long before which time, and a most painful sight it was, the ridge of the shingle roof of his house, a large mansion of but one floor, had a number of these melancholy-looking heralds of death perched thereon . . . the birds must have been directed by smell alone as sight was totally out of the question.[14]

Owen's anatomical evidence for an olfactory sense in vultures was ignored. Other zoologists, contemporaries who dissected the heads of fulmars, albatrosses and kiwis – all of which indicated that these birds have a well-developed sense of smell – were also ignored.[15]

In 1922 John Gurney commented on the curious lack of evidence for a sense of smell in birds when it was well established in other animal groups. He says: 'of the existence of a highly-developed scent in the mammals there can be no shadow of doubt'. In fishes, he writes, it is 'fully acknowledged' that they possess a sense of smell. More remarkably, even certain butterflies and moths are 'credited with the enjoyment of the faculty of scent'. Birds were a puzzle, and olfaction the most perplexing of their senses: 'It is curious that so important a matter should be still unsettled'.[16]

Jerry Pumphrey, by now a professor of zoology at Liverpool, wrote a review of bird senses in *The Ibis* in 1947, and said, after discussing vision and hearing: 'Of the other sense organs, there remains little that is worth saying. The sense of smell is undoubtedly only very moderately developed by comparison with the more gifted mammals.' Pumphrey acknowledged the anecdotal evidence for a sense of smell in some birds, but then pointed out that it was contra-dicted by other anecdotes.[17] Throwing his hands up in despair, he concludes that: 'Indeed, in this field critical experiments are almost impossible, because human beings labour under the excessive difficulty of not knowing what to look for. There is no theory of smell which is even moderately consistent with the facts of human olfactory experience . . .'[18]

A few years earlier Percy Taverner, curator of birds at the National Museum of Canada, had written a short article – a note, really – saying much the same and bemoaning how little was known about the sense of smell in birds: 'These may be difficult subjects but it is time they were tackled. Here is a chance for some ingenious and ambitious post-graduate seeking fame and new worlds to conquer!'[19] Little did Taverner suspect that it would be neither a postgraduate, nor a man, who would launch the scientific study of the sense of smell in birds.

Enter Betsy Bang, a medical illustrator at Johns Hopkins University in the United States in the late 1950s. Single-handedly, she transformed the study of avian olfaction, dragging it out of the academic shadows and into the limelight.

Betsy worked for her academic husband, illustrating his articles on respiratory disease in birds. This meant dissecting and drawing the nasal cavities of various bird species from her husband's extensive anatomical collection. Betsy had only limited training in biology but was a keen amateur ornithologist, and she was smart. As she dissected and drew, she began to wonder why the design of nasal cavities differed so much in different species.

The structures inside the human nose that warm and moisten incoming air, but also detect odours, are called the conchae.[20] The term may be unfamiliar, but the conchae are the wafer-thin leaves of bone inside the harder upper part of the nose that are so easily broken during fights and less easily reshaped during nose jobs. In birds, air is drawn in through the two external nostrils, which in most species are mere slits in the upper portion of the beak. There are three chambers inside the upper beak of most birds; the first two warm and humidify the inhaled air, some of which passes into the lungs via the mouth. The third chamber at the base of the beak contains the conchae, which comprise a scroll-like roll of cartilage or bone. Air passes between the leaves of bone that are covered with a sheet of tissue and in which reside the many tiny cells that detect odours and relay information to the brain. The more complex the conchae – the more turns of the scroll – the greater their surface area and the greater the number of scent-detecting cells. The parts of the brain responsible for interpreting odour lie close to the base of the beak and because of their shape are referred to as the olfactory bulbs.[21]

Looking at her dissections, Betsy simply could not accept that birds with large, complex nasal cavities had no sense of smell, as all the textbooks asserted. She was 'deeply concerned that wrong information was out there about the olfactory abilities of birds and she wanted to correct this misunderstanding'.[22] The reason for the misunderstanding, she guessed, was a lack of communication between anatomists and those conducting behavioural studies. The few recent behavioural studies designed to establish whether birds could detect chemical signals had been conducted on pigeons, a convenient but biologically inappropriate study species which Betsy described as being 'feebly equipped' in the olfactory department. The other problem was that the behavioural experiments themselves were often poorly designed.

For her initial investigation Betsy focused on three unrelated species, each with greatly enlarged nasal conchae but each with a very different lifestyle. They were: (i) the turkey vulture, the species that Audubon thought he had studied, a diurnal carrion feeder; (ii) the black-footed albatross, a pelagic seabird that feeds on squid and whale carcasses (marine carrion), and (iii) the oilbird, a nocturnal, tropical, fruit-eating species that, as we have seen, nests in caves in complete darkness. The anatomical evidence seemed overwhelming – what other purpose could this elaborate nasal tissue have unless it is to detect odours? The resulting paper – her first – was entitled 'Anatomical evidence for olfactory function in some species of birds', and illustrated with slightly ghoulish yet revealing dissections of the heads of each species. The results were published in *Nature* in 1960 and, as one of her colleagues later said, 'Bang's paper made it impossible to deny the existence of a sense of smell in birds'; Betsy made 'an essential contribution at a receptive time'.[23]

Throughout the 1960s Betsy continued to look at the anatomy of different birds, but it was a meeting with Stanley Cobb in the late 1960s that provided the next big step. Betsy and her husband had a second home at Woods Hole at the southern end of Cape Cod, Massachusetts, where they spent each summer. At a dinner party one evening she found herself seated next to Cobb, a retired neuropsychiatrist with a passion for birds and brains. A few years previously Cobb had published a short article on the olfactory bulb in birds. He and Betsy hit it off immediately and joined forces to produce a massive comparative study of olfactory bulb size in the brains of 107 different species.[24]

They measured the length of the olfactory bulb with a ruler, expressing it as a percentage of the maximum length of the brain.[25] They knew that this was a rough-and-ready measure of olfactory potential, but the only way they could have done anything better would have been to dissect out the olfactory bulb, weigh it and then calculate what this represented as a percentage of the remaining brain mass; but this would have been extremely time-consuming (it is a difficult dissection) and it would have meant destroying the museum's specimens. For the time being at least, their simple index did the job.

Here are a few examples, in rank order: the higher the value, the greater the relative olfactory bulb size:

Snow petrel	37
Kiwi	34
Petrel – average	29 (varying from 18 to 33)
Turkey vulture	29
Nightjar – average	24 (varying from 22 to 25)
Hoatzin	24
Rail – average	22 (varying from 12.5 to 26)
Feral pigeon	20
Shorebird – average	16 (varying from 14 to 22)
Domestic fowl	15
Songbird – average	10 (varying from 3 to 18)

Overall, Bang and Cobb's comparative study revealed a twelvefold difference in the relative size of the olfactory lobe across different bird species – from the tiny bulb in the black-capped chickadee (a songbird), to the massive one in the snow petrel.[26] They also assumed that the relative size of the bulb reflects olfactory prowess, a link that was not formally verified until the 1990s when researchers demonstrated an association between bulb size and the threshold for odour detection.[27] Overall, this is what Bang and Cobb were able to conclude: 'Our survey suggests that in kiwis, in the tubenosed marine birds, and in at least one vulture, olfaction is of primary importance, and that most waterbirds, marsh dwellers, and possibly echo-locating species, have a useful olfactory sense. In other species it may be relatively unimportant.'[28]

Inspired by Bang's initial papers, another American researcher, Kenneth Stager, decided to rerun Audubon's behavioural experiments. The anatomical evidence for a well-developed sense of smell in the turkey vulture was convincing, but behavioural evidence was still needed. Stager confronted the problem with gusto, setting up some ambitious field experiments that, among other things, involved blowing air over concealed animal carcasses (and in other cases over nothing, as a control) to see the effect on turkey vultures. The effect was dramatic. The birds could clearly smell the carcass even though they were out of sight. A chance conversation with someone from the Union Oil Company of California resulted in a major breakthrough, enabling him to identify just what it was in the odour of animal carcasses that the vultures homed in on. Stager was told how in the 1930s the company had noticed that leaks in natural gas pipelines attracted turkey vultures. The gas contained ethyl mercaptan (aka ethanethiol), a substance that smells like rotten cabbage (also responsible for the smell of bad breath and flatus); it is also released from decaying organic matter, including animal bodies. Union Oil therefore added higher concentrations of mercaptan to the gas to help them locate leaks. As early as the 1930s the company knew that turkey vultures had a good sense of smell and, sure enough, when Stager blew mercaptan-laden air across the California hills, the vultures came flocking.[29] Not only had he obtained convincing behavioural evidence that turkey vultures use their sense of smell to find food, but he had identified the substance whose odour enabled them do this.

Bang's pioneering anatomical research, together with the comparative study she did with Stanley Cobb on the olfactory bulb, was ground-breaking. But such is the truth-for-now nature of science that before long other scientists started to look at these results with new eyes. Science is always on the move and it was perhaps inevitable that new insights and new techniques would eventually expose the limitations of Bang and Cobb's study. Indeed, this is exactly what Bang and Cobb had done in their own investigation which built on and improved the studies done in the nineteenth century.[30] Bang and Cobb's study was in many ways exemplary science. They measured their specimens as carefully as they could, presented their results clearly but also acknowledged the fact that their estimate of olfactory bulb size was simply an index, modestly hoping that 'these crude olfactory ratios may serve as a guide to its [olfaction's] relative importance'. As we've seen, their main conclusion was that, in addition to the kiwi, tube-nosed marine birds (albatrosses and petrels) and the turkey vulture, 'most waterbirds, marsh-dwellers and waders . . . have a useful olfactory sense'.

During the 1980s there were major improvements in the way comparative studies were conducted. Armed with these new methods, two Oxford scientists, Sue Healy and Tim Guilford, decided to check Bang and Cobb's results. When I asked Sue why she thought this was worth doing, she said that, as well as being interested in the new techniques, she also found the explanation that Bang and Cobb had for the variation in olfactory bulb size rather vague: 'In those days being able to pin down one variable in a comparative analysis was much harder, I guess. Also, I'm a Kiwi and the kiwi has an extraordinarily large proportion of its brain given over to olfaction (and is nocturnal) so it seemed worth seeing if activity played a role in the rest of the variation.' Significantly, she added: 'I have been amazed ever since how little attention is paid to the role of olfaction in bird behaviour, not because people should have noticed our paper but because, once noticed, olfaction seems quite relevant to lots of things birds do.'[31]

There were two main reasons for checking. First, Bang and Cobb had not considered the phenomenon of *allometry* – the way that organs scale with body size. Bang and Cobb implicitly assumed that brain size is directly proportional to body size. It isn't. Larger birds have relatively smaller brains, in exactly the same way that adult humans have relatively smaller brains than babies. When the relative size of organs decreases with body size, this is referred to as negative allometry. Healy and Guilford were concerned that, by ignoring the fact that relative brain size decreases with body size, Bang and Cobb's results might be wrong.[32]

The other thing that Bang and Cobb were unaware of was the

fact that, because many of the species in their comparison were closely related, their conclusions might be biased. Today, this type of bias is called a *phylogenetic* effect (phylogeny is the evolutionary relationship between species), and the way that phylogeny can potentially distort the results of a comparative study like Bang and Cobb's may be seen by considering a different example. In the 1960s two North American ornithologists, Jared Verner and Mary Willson, were looking for explanations for why certain birds had a polygynous mating system (i.e. one male paired with several females). After examining the literature, they concluded that marsh-nesting was the link, suggesting that, because a marsh habitat is highly productive and full of insects, female birds are able to feed their young without the help of the male, thus allowing polygyny to evolve. Since thirteen of the fourteen polygynous North American bird species nested in marshes, the effect of habitat seemed clear-cut.[33] But as later became apparent, there was a snag. Nine of these species belonged to a single family – the Icterids, the North American blackbirds whose ancestor may have been both marsh-nesting and polygynous. In other words, the fourteen species in their sample were not 'independent'; nine shared the same evolutionary history, so the number of comparisons on which they based their conclusion that marsh-nesting is the ecological driver for polygyny was much less than fourteen, and, as a result, much less reliable. It was only in the early 1990s that statistical methods for taking phylogeny into account in such comparative studies became available.[34]

Healy and Guilford's analysis showed that, after accounting for both allometry and phylogeny, the link that Bang and Cobb had found between lifestyle (i.e. living on or near water) and olfactory bulb size disappeared. The lifestyle effect was an artefact because most of the waterbirds came from just a few phylogenetic groups. Instead, Healy and Guilford found that it was mainly nocturnal and crepuscular birds that had relatively large olfactory lobes, consistent with the idea that olfactory prowess develops to compensate for reduced visual efficiency. Not all that surprising, you might think, but it is easy to be smart after the event.[35]

When it was published in 1990, Healy and Guilford's study marked an important advance in our understanding of the ecological factors driving a good sense of smell in birds. But now, twenty years on, it looks like it is about to be overturned, or at least modified, as the truth-for-now process rumbles on. Healy and Guilford did not attempt to improve on Bang and Cobb's simple linear index of relative bulb size – they simply used the original numbers because, without going back to the original specimens and doing a great deal of dissection, it would have been difficult to do otherwise.[36] However, by about 2005 high-resolution scanning and tomography (3-D reconstruction) started to become routine in medicine and biology, making it relatively easy (albeit expensive) to measure accurately the volume of different parts of a bird's brain, including its olfactory bulbs.

Jeremy Corfield and colleagues at the University of Auckland in New Zealand have pioneered the use of 3-D imaging to investigate the structure of birds' brains, and have shown that Bang and Cobb's index is sometimes well off the mark. To be fair, Bang and Cobb knew that this was a possibility and it was for pragmatic reasons that they assumed that, regardless of species, the basic design of birds' brains is similar. The 3-D scanning showed that this is not true. In the kiwi, which was the initial focus of Corfield's work, the brain is unusual in its design: the olfactory lobe is not really a 'bulb' as in other birds, but is actually a flat sheet of tissue covering the foremost part of the brain, and the forebrain itself is unusually elongated. It was because of this that Bang and Cobb obtained a large index for the kiwi, so they got (roughly) the right answer (the kiwi does have a large olfactory region), but for the wrong reason.[37]

The 3-D studies have also revealed anomalies in some other species including the pigeon, whose olfactory bulb turns out to be much larger than anyone imagined,[38] and nicely consistent with its ability to navigate using its sense of smell, as we shall see in the next chapter.

Clearly, using Bang and Cobb's index of olfactory bulb size is risky, and what are needed now are accurate measurements of the *volume* of the olfactory region of the brains of all the birds that Bang and Cobb studied. Given the work that this would entail, it might be some time before such information is available. Meanwhile, researchers have little option but to continue to use Bang and Cobb's original values.

A recent study of the genes involved in olfaction in birds, so-called olfactory receptor genes, used nine species of birds that span the full range of Bang and Cobb's olfactory bulb-size index, and showed that, overall, the total number of olfactory genes is positively associated with olfactory bulb size. In other words, the larger the bulb the more important the sense of smell is likely to be. Two nocturnal species, the kiwi and the kakapo, had the highest number of olfactory genes, 600 and 667 respectively, while the canary and blue tit, as expected on the basis of their relatively small olfactory bulb, had many fewer genes (166 and 218 respectively). There was one anomaly, however: the species with the greatest olfactory bulb size, the snow petrel, had only 212 olfactory genes. It is just possible that a 3-D scan might reveal this species' bulb to be not as large as Bang and Cobb suggest, or possibly the snow petrel, which is diurnal, may be sensitive only to a limited range of odours and therefore require fewer genes.[39]

Apart from the publication of Jane Austen's *Pride and Prejudice* and the ongoing Napoleonic Wars, the most significant event of 1813 was Europe's discovery of the kiwi. George Shaw, keeper of zoology at the British Museum, was given an incomplete skin – now known to be a South Island brown kiwi – by Captain Barclay, the captain of a convict ship. Barclay must have obtained the specimen from someone else, for he never visited New Zealand. Shaw described and illustrated this remarkable bird in 1813, naming it *Apteryx australis* (wingless southerner). On Shaw's death later that year, the specimen passed into the hands of Lord Stanley, 13th Earl of Derby, whose enormous collection of natural history specimens at Knowsley Park in turn ended up in the nearby Liverpool Museum, where it has been ever since.[40]

Despite its extraordinary appearance and incomplete nature, Shaw perceptively recognised that the kiwi might be a distant relative of the ostrich and emu (the ratites). Others erroneously imagined it to be either a kind of penguin or a species of dodo.[41]

For over a decade Shaw's was the only kiwi specimen available and some began to doubt the bird's very existence. In 1825, Jules Dumont d'Urville provided some tantalising new information. Recently returned from New Zealand, he described an encounter with a Maori chief wearing a cloak made of kiwi feathers. Following an appeal for more information, some New Zealand settlers put pen to paper and provided the first descriptions of kiwi behaviour, while others sent actual specimens. Once again Lord Stanley was a key player, passing the specimens on to Richard Owen at the British Museum, who, in his meticulous fashion, undertook detailed dissections. Owen noted the uniquely positioned nostrils at the bill tip and from the structure of the brain case recognised that a sense of smell might be important: 'In the interior of the cranium the olfactory depressions are seen to be proportionately larger than in other birds' and those cavities which in other birds are devoted to the lodgement of the eyes, are here almost exclusively occupied by the nose.' Wrapping up, Owen concluded presciently: 'The sense of smell must be proportionately acute and important in the economy [lifestyle] of the Apteryx.'[42]

Observations of kiwis both in the wild in their native New Zealand and among birds brought to Britain and maintained in captivity revealed that they forage by snuffling around, literally, usually in the undergrowth, probing their long bills into the ground in search of their invertebrate prey – mainly earthworms. In the 1860s the kiwi's manner of finding food was accurately illustrated in a series of beautiful watercolour images by the Reverend Richard Laishley.[43]

That kiwis regularly blundered into things when running away from human observers confirmed that their eyesight was poor, but the fact that they audibly snuffled as they foraged strongly suggested that kiwis found their prey by smell. Then, in the early 1900s, W. B. Bentham of Otago University Museum, Dunedin, who knew of the kiwi's large olfactory lobes from Owen's publications, decided to see just how good its sense of smell was. Accordingly, he asked Mr Richard Henry, the curator of Resolution Island, a bird sanctuary off the south-west of New Zealand's South Island, to conduct some simple experiments on a (tame) kiwi, which he referred to by its Maori name, *roa-roa* – meaning 'long' and presumably referring to its beak.

Following Bentham's instructions, Henry presented the kiwi with a bucket that either did or did not contain earthworms buried beneath a layer of soil. The bird had no problem telling where the food was: 'When I put down a bucket of earth without worms in it, the bird would not even try it; but the moment a bucket containing worms was put down the roa was full of interest and commenced to probe at once with its long beak.' Bentham excuses himself for not having undertaken these experiments personally, pointing out the inaccessibility of Resolution Island and how 'the uncertainty of getting back to the mainland in any reasonable time was so great that I had to give up the idea'. Admitting that many other experiments still needed to be done, he felt that his results afford 'a certain amount of evidence for the existence in Apteryx of a keen sense of smell'.[44]

In 1950 Bernice Wenzel joined the faculty in the School of Medicine at the University of California, Los Angeles. She had previously completed a PhD at Columbia University on the sensitivity of humans to smell, but by the time she reached California her research had shifted to the study of brains and behaviour. Even though she had changed direction, a colleague invited her to give a lecture at an olfaction conference in Japan in 1962. She declined, pointing out that she was no longer studying olfaction. Refusing to take no for an answer, her colleague told her she'd 'think of something' and added her to the list of speakers. Bernice began to wonder what she might do and decided to see how pigeons, which she had in the lab, would respond to smell. Using a method commonly used by physiologists, she checked whether the pigeon's heart rate changed in response to different stimuli. Bernice's test involved exposing birds to a stream of pure air interspersed with brief periods during which an odour was added, and the bird's heart and breathing rate were measured. On her very first test Bernice was amazed to see the bird's heart rate soar as the odour was added. Here was unequivocal evidence that the pigeon had detected the smell. More studies quickly followed and at the meeting in Japan she presented her first paper on the sense of smell in birds.[45]

One of just a handful of female professors of physiology in the United States in the 1960s, Bernice Wenzel's great strength was to employ the combined tools and ideas of anatomy, physiology and behaviour better to understand olfaction. Examining birds as diverse as canaries, quail and penguins, she found that every species, including those with the tiniest olfactory lobes, was able to detect odours. Although all species responded, those with larger olfactory lobes showed a greater increase in heart rate. Despite these remarkable results, it was not known whether birds (other than the kiwi) used olfactory information in their daily lives.

The heart rate experiments were so successful that Wenzel decided to try the same approach with kiwis. In her previous investigations the birds simply needed to have their wings restrained for them to sit quietly during the experiment. Not kiwis. They are immensely powerful birds, and she quickly discovered that, with virtually no wings to constrain them and very strong legs, adult kiwis were 'able to wiggle out of almost any system of restraint'. Instead, Wenzel took readings from a single young kiwi that was accustomed to being handled. To validate her results, she obtained a few readings from an adult – much more aggressive – bird.[46]

Curiously, and in contrast to all the other birds Wenzel had assessed, odour caused no change in the young kiwi's heart rate, even when given a whiff of its favourite earthworms. Instead, it was changes in breathing rate and alertness that most obviously revealed the bird's ability to detect odour. Wenzel then carried out some behavioural experiments to see if kiwis (five different individuals) could detect food by smell alone.

Employing an experimental design very similar to the one used by Bentham and Henry fifty years earlier, she presented the birds with metal tubes sunk into the ground. Some tubes contained strips of meat that the birds were used to eating, but covered with a layer of damp soil, while the other tubes contained only damp soil. In both cases the tubes were covered by a thin nylon mesh that the bird had to puncture with its bill so as to access the soil. This was essential because the birds foraged only at night and it would have been difficult or impossible to see which tubes they probed. The metal tubes provided no other cues to the presence of food and because the meat was inert it created no sound; there were no visual cues because the appearance of all the mesh-covered tubes was identical. The mesh covering also precluded any taste cues and left convenient evidence if the bird pushed its beak through it.

Just as in the earlier studies, the kiwis were interested only in the tubes containing food. Not only that, but they probed directly on to the food items, indicating that they could detect extremely subtle gradients in odour.

Other behaviours in her captive kiwis suggested to Wenzel a strong reliance on smell. One night while she was in the aviary, a kiwi woke up early and approached her. In her own words: 'It was dark, the bird stopped very close to me and then methodically moved the end of its beak up and down my legs without actually touching them, as if outlining me . . . the behaviour is far more consistent with dependence on olfaction than vision.'[47]

Almost everyone who has watched free-ranging kiwis has commented on their audible sniffing, but it is generally recognised that this is as much to do with clearing the nostrils as with olfaction. Kiwis have nasal glands that secrete mucus when they are excited (by food) and, because the external nostrils are narrow slits, they become easily clogged with soil during probing.

During her studies, Wenzel noticed that lightly touching the bill tip of one of her captive kiwis would result in active searching movements, indicating that touch was also an important part of natural foraging behaviour. And in concluding her account she said: 'there is probably a close interaction between the tactile and olfactory modalities, with little, if any, visual participation'.[48]

The northern hemisphere's counterpart to the kiwi is the woodcock. Apart from its enormous eyes – which are essential for its crepuscular activity and flight, including nocturnal migration – the woodcock and kiwi are very similar. Both species have a similar lifestyle, probing for worms beneath the soil surface. As long ago as 1600, Ulysses Aldrovandi tells us in his encyclopaedia of birds that woodcocks find their food by smell. This seems to have been well established since he quotes a poem on bird-catching by Marcus Aurelius Nemianus dating to AD 280, which mentions the bird's enormous nostrils and its capacity to smell worms. Several later authors also refer to the woodcock's sense of smell, but curiously, and in contrast to many other ornithological facts, they do so without citing or plagiarising earlier writers, suggesting that the woodcock's sense of smell may have been independently discovered several times. Buffon, for example, quotes William Bowles, who in his book of 1775, *An Introduction to the Natural History and Physical Geography of Spain*, describes how he watched a woodcock in the royal aviaries probing for worms in damp soil: 'I did not see it once miss its aim: for this reason, and because it never plunged its bill up to the orifice of the nostrils, I concluded that smell is what directs it in search of its food.' And then, citing his colleague René-Joseph Hébert, a hunter and naturalist, Buffon adds this: 'But nature has given, at the extremity of its bill, an additional organ, appropriated to its mode of life; the tip is rather flesh than horn, and appears susceptible to a sort of touch, calculated for detecting prey in the mire.'[49]

The English ornithologist George Montagu, who dissected many woodcocks and observed one living in his aviaries in the late 1700s, writes:

Thus when most other land birds are recruiting exhausted nature by sleep, these [woodcock] are rambling through the dark; directed by an exquisite sense of smelling, to those places most likely to produce their natural sustenance; and by a still more exquisite sense of feeling in their long bill, collect their food. . . . The nerves in the bill . . . are numerous, and highly sensible of discrimination by the touch.[50]

Writing about the senses of birds a century later, John Gurney had this to say:

> The investigator has to be cautious not to confuse the organ of scent with that of touch, by means of which some birds feed – e.g. the woodcock. Thus it will be seen what an involved business it is for an experimenter to formulate any trial which appeals to a bird's sense of smell, and which at the same time excludes sight, hearing and touch.[51]

When I checked, I was surprised to see that Bang and Cobb[52] had obtained an olfactory bulb index of just fifteen for woodcock, which places it in the mid-range rather than close to the top. I wonder if this means that the olfactory bulb is an odd shape, as 3-D scanning revealed for the kiwi, and that the index is wrong: given the wood-cock's unusually shaped skull, this is a possibility. Of course, the other thing to do is conduct some behavioural studies and put the woodcock through its olfactory paces to see how it compares with the kiwi.

Bernice Wenzel and Betsy Bang put avian olfaction on the academic map, partly through their independent research but also through a chapter of a book they wrote together in the 1970s that has become the definitive account of the sense of smell in birds.[53] Betsy Bang died at Woods Hole in 2003 at the age of ninety-one, and Bernice, now in her eighties, is an emeritus professor at UCLA. In 2009, two other female pioneers of avian olfaction, Gaby Nevitt and Julie Hagelin, dedicated a symposium to their two predecessors. Bernice told me that she was overwhelmed by this gesture, and remarked on how different it was to the early comments on her research, many of which had questioned why she was even bother-ing to study smell in birds.[54]

What was it about the field of avian olfaction, that it should have been so dominated by women? Few other areas of research – except for primate behaviour – have such a preponderance of female researchers. Colleagues I have spoken to told me that as mentors Betsy and Bernice were extremely encouraging and more generous in sharing advice than most male researchers would have been, traits that may have been particularly appealing to younger female zoologists.

In 1980, with colleagues Richard Elliot and Remey O'Dense, I visited a remote and little-known group of islands, called the Gannet Clusters, some twenty miles off the coast of Labrador. Our goal was to count the seabirds there. This was hardly a trivial task for there were tens of thousands of puffins and guillemots, and only slightly fewer razorbills, plus a few fulmars and kitti-wakes (but no gannets – the name of the islands is misleading and its origin a mystery). On our first night, not long after we had settled down to sleep inside our tent, Richard sat bolt upright exclaiming, 'Leach's petrel!'

I woke and listened, and, sure enough, outside in the dark I could hear the distinctive gentle purring of a Leach's storm petrel close by. The reason Richard was so excited was that this was the first record for this tiny, nocturnal seabird on these islands and one of the most northerly records for North America. The next morning we hunted around outside the tent for further signs, and there in the peaty soil was a nesting burrow, just five centimetres in diameter. Richard's immediate reaction was to drop to his knees, stick his nose into the hole and sniff audibly. 'Yes!' he said. 'It's Leach's all right', for, like other members of the petrel family (which includes the albatrosses and shearwaters), Leach's petrel has a distinctive musky smell.

Continuing to search, we found several more burrows, and, as luck would have it, inside one of them I found a mummified Leach's petrel corpse, definitive evidence of their existence. In a somewhat macabre but entirely scientific gesture, I kept the dead bird: it was completely dried out and not in the least bit unpleasant. Years later, back in my office in Sheffield, I had only to sniff the bird to be transported back to the magic of the Gannet Clusters, so strong and so evocative was the bird's aroma.

Bang and Cobb had not included Leach's petrel in their comparative study, but they examined ten other petrel species, all but one of which had huge olfactory bulbs. Indeed, ever since the early days of commercial whaling, mariners had noticed how incredibly sensitive albatrosses, petrels and shearwaters appeared to be to the odour of whale offal. In the 1940s, Loye Miller, professor of biology at the University of California, Los Angeles, conducted some simple but extremely telling experiments involving individually marked black-footed albatrosses – which he refers to as 'goonies' – off the west coast of North America.[55] Within one hour of bacon fat being poured on to the sea surface, birds were drawn in – Miller estimated from a distance of 32 km. No birds were attracted to paint scum, an equally smelly substance used as a 'control'. 'Chumming' is now used regularly by oceanic twitchers to attract seabirds, the olfactory equivalent of playing a recording of birdsong to attract land birds. The effect is remarkable, as I witnessed off the east coast of New Zealand's South Island, at Kaikoura: being surrounded by fifteen different species of petrels and albatross just a few metres away counts as one of my best birdwatching experiences.[56]

Scientists refer to albatrosses, petrels and shearwaters as tubenoses. Despite their obvious link with odour detection, the function of their tube-like nostrils remains a mystery. Different species, which range in size from the 50-g storm petrels to 8-kg wandering albatross, feed on krill and squid, and sometimes on whale offal. Finding a decomposing whale carcass by smell might not be all that difficult – the aroma of rotting blubber is one that can stick in human nostrils for hours or days, as I can verify – and even we might not find it too difficult to travel upwind to such a feast. But krill and squid – do they smell strongly enough to allow tubenoses to find them in the vast, featureless ocean? That's another story.

Gaby Nevitt, mentioned earlier, and a biologist at the University of California at Davis, started out studying how salmon relocate their spawning river after several years at sea. The idea that they might use smell to navigate once seemed preposterous, yet research in the 1950s showed it to be true.[57] Almost as unbelievable is the way albatrosses, flying across vast tracts of the ocean, are able to relocate their breeding colonies, tiny specks of rock in a featureless sea. There is no question that they can do so, but what wasn't appreciated until the 1990s was just how far they travelled from their colonies in search of food during the breeding season. Some wonderful pioneering work by French researchers Pierre Jouventin and Henri Weimerskirch showed, by means of what was then new satellite tracking technology, how wandering albatrosses covered thousands of kilometres in search of food, and still unerringly managed to find their breeding island.[58] Gaby became interested in *how* albatrosses were able to find food and relocate their colonies so efficiently.

Smell seemed a likely candidate, mainly because of the abundant anecdotes of whalers, fishermen and birdwatchers. In addition, studies by Tom Grubb, a PhD student at the University of Wisconsin (and later at Ohio State University) in the 1970s, showed that Leach's storm petrels – the same species we had discovered in Labrador – invariably returned *upwind* to their breeding islands in the Bay of Fundy. Much more significantly, Tom, working with

Betsy Bang, showed how petrels whose olfactory nerve had been cut (an operation that renders birds anosmatic – smell-blind) were unable to relocate their colony, whereas unoperated-upon birds could do so, and from as far away as Europe.[59]

Smell was clearly important in allowing Leach's petrels to relocate their nesting colony. But this was only half the story. Gaby Nevitt was interested in whether smell also played a role in their finding food. She started by repeating the kinds of experiments Loye Miller and others had carried out, pouring smelly slicks on to the ocean and seeing how quickly birds were attracted, compared with their attraction to some other, unsmelly, substance. In 1980, Bernice Wenzel's graduate student Larry Hutchinson had shown that ground-up krill poured on to the ocean attracted sooty shearwaters, indicating that something in the krill pulled in the birds. As Nevitt soon found out, conducting experiments on oceans where 12-m waves are the norm was far from easy. She used vegetable oil laced with raw krill extract and unadulterated vegetable oil as a control. The studies confirmed that the smell attracted birds like petrels and albatrosses very effectively, but it didn't really answer the question of whether krill give off a particular odour that helps the birds locate them.[60]

Then, in 1992, in unusual circumstances, Gaby met Tim Bates, an atmospheric scientist. In her own words:

I was doing a cruise down near Elephant Island (off Antarctica) and we ran into some very bad weather . . . I got thrown into a tool chest in a storm and injured my left kidney. Of course I didn't know that at the time but the pain was so bad that I was confined to my bunk which was down in the bowels of the ship. We were within a week of getting into Punta Arenas and I swear it was the longest week of my life. Anyway, when we got there, I was not very mobile. The new chief scientist was Tim Bates, and he was kind enough to let me stay aboard to wait for transport home. During this time his team was outfitting the ship for their atmospheric cruise on dimethyl sulfide (DMS).[61]

DMS is a biogenic substance released from the bodies of phytoplankton when they are eaten by zooplankton such as krill. DMS dissolves in seawater and is then released into the atmosphere, where it lingers for hours or even days.

Gaby continued:

Once I was able to see some of their transect data and smell DMS and get some pain killers the world changed. The profiles he showed me were like mountain ranges or landscapes. DMS was just one tractable compound, but it suddenly seemed that 'tracking the ephemeral plume to the prey patch' was the wrong model for large-scale questions. Instead, the ocean was overlain with odour landscapes tied in part to bathymetric features, shelf breaks, sea mounts, etc, and that changed my thinking entirely. When I think back on it, if I hadn't had such a bad accident, I wouldn't have met Tim, and I would probably still be chumming fish guts without seeing the bigger framework.[62]

A stream of experiments followed, including one showing that even at their breeding colony (rather than out at sea) Leach's petrels were attracted to DMS. A study of Antarctic prions – another petrel species – showed that they were attracted to artificially created DMS-laden slicks at sea. Particularly revealing was an experiment that was actually a rerun of one performed by Wenzel in her early research, which involved measuring changes in heart rate in response to specific odours. Working on Ile Verte in the Kerguelen Archipelago, in the southern Indian Ocean, Antarctic prions were gently removed from their breeding burrows and taken to a nearby temporary laboratory. Electrodes were carefully (and temporarily) attached to the skin, allowing Nevitt and her colleague Francesco Bonadonna to measure the birds' heart rate on an electrocardiograph, as air, with or without DMS, was passed over the birds' nostrils. The crucial part of this study was that the concentrations of DMS experienced by the birds during these brief experiments were similar to those they would experience out at sea. In response to pure air, none of the birds showed any change in heart rate, but in response to DMS all ten birds exhibited a pronounced increase, thus providing some of the best evidence so far that naturally occurring odours may help birds like prions navigate across the ocean.[63]

Nevitt began to wonder whether substances like DMS might provide oceanic seabirds with an olfactory landscape, or rather, an olfactory seascape, superimposed on the surface of the ocean. Areas where phytoplankton accumulate, such as fronts and upwellings, attract predatory zooplankton like krill. As the krill consume the phytoplankton, DMS is liberated into the air, creating a plume of odour downwind from the source. Wind and wave action will render the plume patchy and irregular and, of course, weaker and weaker the further it is from the source. How might we expect a bird to behave if it was using such airborne information to find prey, the source of the odour plume? The answer is to fly crosswind to maximise the chances of locating a plume and, once it has been detected, to fly upwind in a zigzag manner – casting from side to side – to retain contact with the odour trail until it finds the prey.

The match between Nevitt's prediction and some early observations of foraging petrels is striking. In his account of how New

England fishermen caught petrels to use as bait, Captain J. W. Collins wrote this in 1882:

> On many occasions during the prevalence of dense fog, when not a bird of any kind has been seen for hours, I have thrown out as an experiment, pieces of liver to ascertain if any birds could be attracted to the side of the vessel. As the particles of liver floated away, going slowly astern of the schooner, only a short time would pass before either a Mother-Carey Chicken [storm petrel] or a Hag [hagdon, hag-down or greater shearwater] . . . could be seen coming up from the leeward [upwind] out of the fog, flying backward and forward across the vessel's wake, seemingly working up the scent until the floating pieces of liver were reached.[64]

To test her ideas, Gaby Nevitt and colleagues employed some stunning new technology on the world's largest seabird, the wandering albatross. This species forages over thousands of square kilometres in search of its squid or carrion prey and, like other tubenoses, has an exceptionally large olfactory bulb. It is also known to be attracted to fishy odours, making it a prime candidate for a study of odour detection. Nineteen wandering albatrosses, rearing their chicks on Possession Island in the Southern Indian Ocean, were fitted with GPS (global positioning system) locators that allowed the researchers to follow with extraordinary precision the birds' oceanic flight paths prior to the capture of prey. The birds were also fitted with a stomach temperature recorder that detects when a bird has eaten something.

If the albatrosses were foraging by sight, it was predicted that they would fly pretty much in a straight line towards their prey, but if they were using odour they should adopt a zigzag flight path. In fact, about half of all feeding events involved zigzag flights, suggesting that these albatrosses use odour plumes about half the time when finding prey. This remarkable study provides further convincing evidence that olfaction plays a fundamental role in the albatross's foraging, but, just as in other species, olfaction is used in conjunction with other senses, in this case, vision.[65]

The idea of an olfactory seascape is relatively new; the idea of an olfactory landscape is not. In the 1970s, prior to the start of Gaby Nevitt's career, Italian researchers led by Floriano Papi suggested that pigeons used olfaction as part of their repertoire of navigational abilities. In contrast to Gaby Nevitt's olfactory seascape, the idea that pigeons use olfactory cues to facilitate their homing abilities has had a rocky ride. Part of the difficulty has been in disentangling the role of olfaction from the ability to sense the earth's magnetic field. Making the pigeon problem even more intractable is the nerve (the ophthalmic branch of the trigeminal nerve (VI)) that connects to putative magnetoreceptors in the upper part of the beak.[66] Because it is extremely difficult to cut the olfactory nerve without also cutting this nerve, most previous experiments cut both, thereby 'knocking out' both senses. Recent work by Anna Gagliardo, at the University of Pisa, in Italy, however, has dealt with this problem, and concludes that olfactory cues are indeed necessary for the development of the navigational map in pigeons.

Let's finish this chapter by returning to João dos Santos's honeyguides. Kenneth Stager – who corrected Audubon's erroneous conclusions about olfaction in turkey vultures – conducted his own simple honeyguide experiment in the 1960s. During fieldwork in an area of Kenya where honeyguides were common, Stager placed a pure beeswax candle in the crotch of a tree. Unlit – he doesn't say for how long – the candle attracted no honeyguides, but within fifteen minutes of lighting it a single lesser honeyguide had appeared, and after thirty-five minutes there were no fewer than six honeyguides near the candle or nibbling the soft melted wax. Stager took his study one step further and collected 'cranial material of three [honeyguide] species'. His subsequent dissections confirmed that all three species have exceptionally large olfactory conchae, which, as he said, strengthened 'the belief that olfaction may well play an important role in the behaviour of honeyguides'.[67]

6

Magnetic Sense

Bar-tailed godwits on migration. Guided by a magnetic sense, these birds fly from Alaska to New Zealand in a single, non-stop, eight-day, 11,000 km flight.

A faculty sometimes hypothetically invoked, but not known to exist.
Arthur Landsborough Thomson entry for 'Magnetic sense', 1964,
in *A New Dictionary of Birds*, Thomas Nelson & Sons

I am on Skomer Island climbing carefully down a steep, rocky slope towards an unsuspecting group of guillemots. Most of the birds are brooding a single chick, each of which, I like to think, is imagining where its next meal might be coming from. Far below, the waves are crashing on to the black basalt rocks, and away to the east under a clear blue sky I can see the hazy outline of the wild Pembrokeshire coast. I stop just above a group of guillemots and edge forward with my modified fishing pole. After making myself secure, I carefully hook one of the adult birds round its leg. As I draw the bird towards me, it is a few moments before the bird is aware that something is amiss. But too late! Before it realises what is happening, I have the guillemot firmly in my grasp. This apparent stupidity, tameness or lack of awareness was what in the past gave this species the name 'foolish guillemot'. Luckily for me the birds *are* a touch naive and, one after another, over the next hour I capture a total of eighteen birds. As each one is caught we place a metal ring (band) on one of its legs, and on the other leg we place a specially modified plastic ring which carries a tiny device, a geolocator, that will record the amount of daylight every ten minutes, until the battery runs out in two or three years' time. The amount of light at different latitude and longitude varies, allowing us to establish where the bird has been. As soon as each device is attached we release the birds into the air: they hurtle out to sea, describe a large arc and, a few minutes later, with a flurry of wings, are back on the ledge and reunited with their chick.

I have been studying guillemots on this island since the 1970s and, as I write, it is now 2009. I'm working alongside Tim Guilford and his students from Oxford, and my long-time colleague from Sheffield, Ben Hatchwell, who also studied Skomer's guillemots for his PhD research.

Twelve months on and I'm roped up again for the descent to the same small group of guillemots. This time it is different: once caught, twice shy. The guillemots know what to expect and, despite being emotionally tied to their chick, are determined not to fall prey to my hook again. It is me who is beginning to look foolish, for my colleagues and I are desperate to recover the geolocators so that we can see where our birds have been over the past year. Little is known about where Skomer's guillemots spend their winter other than what has been gleaned from the recovery of ringed birds found dead – a crude and possibly biased picture.

Perched 70 m above the sea, I lean forward with my guillemot hook, my arms stretched to the limit, the birds edging further away, dancing around the hook, determined not to get caught. After thirty minutes, I give up and climb back up the rope to where, out of sight of the birds, my expectant colleagues are waiting. They are disappointed at my failure, as am I, and Ben offers to have a go.

He disappears over the edge until all we can see is the top of his climbing helmet, and occasionally the end of the pole, as he patiently edges closer to the birds. When the birds are on high alert like this, the only hope is that there is some kind of distraction, like a fight or a bird arriving from the sea with a fish. That's exactly what happens – a fight (I can hear the aggressive calls) – and I see Ben move the pole in a purposeful manner. Suddenly he's up the rope and, with a huge grin, hands me a guillemot bearing a distinctive green geolocator.

We carry the bird another 70 m up to the cliff top where Guilford's students are waiting. The geolocator – still on the bird's leg – is plugged into the laptop and the data downloaded. Satisfaction isn't guaranteed: sometimes the devices fail. But not this one. Within minutes of capture, like magic the bird's previous 370 days appear on the computer screen. Lying on the grass, we cluster around the laptop, shading the screen from the sun. A map of the world appears, pinpointing every ten-minute fix, until the bird's entire year of travel emerges.

This is what we see: soon after the end of the previous breeding season last July, the bird set off south for the Bay of Biscay, spending a few weeks there before flying 1,500 km north to spend much of the winter off north-west Scotland. Then, back down to Biscay again in the weeks before the start of the current breeding season and then back on to this very ledge on Skomer.

This is instant gratification, a year's worth of unique data delivered in a few moments on a computer screen. It seems miraculous, and, indeed, the new tracking technology – geolocators, satellite trackers and so on – has resulted in a revolution in the study of bird movements, migration and navigation.

We later recover the geolocators from several other guillemots and, reassuringly, they all show similar patterns, providing us with a dynamic picture of the huge distances these birds move during the winter months while away from the colony.

This is novel guillemot information. These few results completely change our view of their movements based on decades of ringing recoveries. Ben and I are delighted; for years we had fantasised about where our guillemots might go outside the breeding season. But, compared with some of the recently completed studies on other species, ours is a modest success. Geolocators have recently been used on birds as small as red-backed shrikes and nightingales, tracking their migratory movements from northern Europe to Africa and back. In terms of distances travelled, however, the most spectacular results come from shearwaters, albatrosses and Arctic terns, all of which undergo monumental oceanic journeys, and particularly impressive is the bar-tailed godwit's eight-day, 11,000-km non-stop flight from Alaska to New Zealand.[1]

As we sit on the Skomer cliff top in the sun, the map on the computer screen in front of us raises an important question. How, with nothing but an oceanic horizon, do guillemots know which way to fly to find their breeding colony, or, indeed, to find their feeding areas in Biscay or off northern Scotland? How do those godwits traversing the entire Pacific Ocean know where to go? How birds find their way – not just during migration, but during their everyday lives – is a question that has been asked many times in the past millennia.

Two of the many people who asked this same question were David Lack and Ronald Lockley in the 1930s. Lack was then a schoolteacher at Dartington in Devon, studying robins in his spare time, later famous for his book *The Life of the Robin* (1945), and somewhat later as the most renowned ornithologist ever. Ronald Lockley was an amateur ornithologist, who, in 1927, at the age of twenty-six, set up home with his wife Doris on the uninhabited island of Skokholm, five kilometres to the south of Skomer. Over the next few years Lockely studied the island's seabirds, including its most numerous and mysterious species, the Manx shearwater. Around 150,000 pairs of shearwaters breed on the adjacent islands of Skomer and Skokholm, some 40 per cent of the world population. The birds are nocturnal, to avoid predatory gulls, and come ashore between March and September only to breed, spending the rest of the year at sea. Lockley's investigations of the bird's breeding biology broke new ground for, at that time, very few seabirds had been studied in any detail.

In June 1936, Lack took a group of schoolchildren to Skokholm where they camped alongside Lockley's tiny whitewashed cottage. One evening, as dusk fell, Lack and Lockley began to talk about how birds found their way, speculating about what might happen if Lack were to take a shearwater back to Devon with him: how quickly would it return to Skokholm? The children listening in loved the idea so when Lack and the children left Skokholm on 17 June they took with them three shearwaters, each bearing a unique ring. Sadly, two birds died en route, but the third, a bird that Lockley had named Caroline (he was unashamedly anthropomorphic about his ringed birds), was duly released from Start Point, southern Devon, at 2 p.m. on 18 June, some 225 miles (360 km) by sea from Skokholm. Communication between Devon and Skokholm was limited to the postal service, which could take several days, so Lockley was unaware that two of his precious birds had died. Not expecting any bird to return until 19 June at the earliest, Lockley nonetheless decided to check their burrows just before midnight on the evening of 18 June. To his amazement Caroline had returned and was incubating her egg, just nine hours and forty-five minutes after release. Ecstatic, Lockley wrote: 'It is clear . . . that Caroline knew the way. She had no time for searching. She recognised in what direction Skokholm lay and made for it. Our success with Caroline was provocative. It suggested further experiment.'[2]

To establish whether the shearwaters had a true navigational sense, Lockley and Lack realised that they had to be released from locations they could not possibly have visited previously. Accordingly, there followed a succession of 'releases' from a variety of increasingly ambitious locations, including an inland site in Surrey in the UK, from Venice in Italy and from Boston in the USA. The rapidity with which some of the birds returned to Skokholm reaffirmed a strong navigational sense.[3]

Lockley's pioneering studies were continued and developed by Geoffrey Matthews, an ornithologist at the Wildfowl Trust in Slimbridge, Gloucestershire, in the UK, who elevated the operation to a rather more scientific level in the early 1950s. Matthews released birds from a variety of locations, including the top of the Cambridge University library tower, noting their direction of departure and being careful not to release the next bird until the previous one had flown out of sight (to avoid the possibility of them influencing each other). The majority of birds left the library tower in a westerly direction and, flying across country, returned directly to Skokholm, 'providing the first unequivocal evidence of a true navigational ability in a wild bird'.[4] I wonder what any birdwatcher, oblivious to the great experiment, might have thought had he chanced to see one of these shearwaters winging its way westwards so many miles from the sea.

Not only did Lockley start the shearwater navigation studies, but he was the first to examine the bird's breeding biology, establishing that its incubation period was fifty-one days; the fact that partners incubated in alternating six-day shifts; and that the slow-growing chick spent no less than ten weeks in the burrow before fledging. Lockley left Skokholm in 1939 just before the outbreak of the Second World War, but in the early 1960s there was renewed inter-est in Skokholm's shearwaters when Mike Harris started his PhD studies there. Harris began ringing large numbers of shearwater chicks in an effort to understand better the birds' biology, and between 1963 and 1976 ringed – with the help of what were affec-tionately known as the 'shearwater slaves' – a staggering 86,000 birds. A fortuitous spin-off from all this ringing was a number of recoveries that provided a glimpse of where the shearwaters went outside the breeding season. It was already known that Manx shear-waters occasionally appeared in the southern hemisphere; the great seabird biologist Robert Cushman Murphy had seen one off the coast of Uruguay in 1912, but it was assumed that such sightings were of birds whose breeding colonies were at the very south of their range in the Azores. Confirmation that Skokholm birds some-times travelled the 10,000 km to South America first came from a ringed bird found dead on the coast of Argentina in 1952. But, of course, one swallow (or in this case one shearwater) makes neither a summer nor a convincing case for regular long-distance movements.

That Manx shearwaters do indeed regularly winter off the coast of South America was elegantly confirmed in the 1980s when Mike Brooke and his former PhD supervisor Chris Perrins decided to look at the 3,600 recoveries of ringed shearwaters accumulated over the previous twenty years. Using ringing recoveries to infer movement patterns of seabirds is a bit like trying to ascertain the summer holiday locations of British tourists from the police stations where their lost passports are handed in – crude at best and subject to all sorts of biases. The recoveries suggested that the shearwaters leave their breeding colonies on Skokholm and elsewhere in Britain in the autumn, fly south past the Bay of Biscay, on past Madeira, the Canary Islands and West Africa, and then somewhere near the equator cross to South America, arriving off the coast of Brazil. The return journey the following spring starts with birds heading out into the central South Atlantic and then heading back to Britain via a slightly more westerly route than their southward migration.[5]

In August 2006, Tim Guilford and his colleagues placed geolocators on the males and females of six pairs of Manx shearwaters breeding on Skomer Island. Because shearwaters nest in burrows, they are much easier to recapture than guillemots. The following spring, soon after the female had laid her single egg, all twelve birds were recaptured. The geolocator analysis confirmed the broad pattern of movement previously deduced from fifty years of accumulated ringing recoveries, but provided some unexpected information as well. First, the birds wintered further south than ringing recoveries suggested: off the coast of Argentina, south of the Rio del Plata in an area of mixed ocean currents that presumably provides rich fishing for the birds. Second, it was previously thought, on the basis of the occasional very rapid ringing recovery

– including a bird that was found on the coast of Brazil just sixteen days after ringing – that shearwaters fly directly to their wintering grounds. The geolocator information showed that such rapid, direct flights are not typical: rather, the birds have frequent stopovers, much as terrestrial migrants do, presumably to refuel. In some cases, shearwaters remained at their stopover locations for a couple of weeks.[6]

While this new technology has extended and refined our view of the huge global distances some birds travel, so far at least it has not provided many new insights into *how* birds make these journeys and how they find their way.

Paradoxically, perhaps, it has been through the study of *captive* birds that we have gained most understanding of navigational mechanisms. In the early 1700s casual observers of caged songbirds like the nightingale noted how their birds began an agitated hopping each autumn and spring when they would normally be migrating. Two hundred and fifty years later, in the 1960s, biologists were finally able to capitalise on this so-called migratory restlessness through the use of an ingenious device known as an Emlen funnel, after Steve Emlen who invented it.[7]

The Emlen funnel revolutionised the study of bird migration. It consists of a blotting paper funnel about 40 cm in diameter at its widest, with an ink pad at the bottom, and a domed wire mesh top – through which the birds can see the sky. As the bird hops, the ink on its feet leaves a trace on the blotting paper which provides an index of both the direction and the intensity of migration.[8] The beauty of the Emlen funnel is that it is cheap and allows researchers to test large numbers of (small) birds very quickly. Sometimes it is only necessary to place a migrant in the funnel for an hour or so to obtain a meaningful trace. Using this method, which has been verified in many different ways, we now know that small birds have a genetic programme to fly in a particular direction for a certain number of days. Although this is remarkable, on its own it is insufficient to tell us how birds navigate. Certainly, it cannot explain how a Manx shearwater on a featureless Atlantic Ocean knows how to return to Skomer or how a nightingale, resting at an oasis in the Sahara on its spring journey north, knows how to find the previous year's territory in a Surrey woodland.

The study of how birds find their way has had a long and sometimes acrimonious history. In the mid-1800s, there were two main ideas for how birds like pigeons found their way home. One was that birds remembered their outward journey, a notion for which there is no evidence. The other idea was based on the relatively recent discovery that the earth behaves like a gigantic magnet and that birds possess a sixth sense, allowing them to detect the earth's magnetic field. The novelist Jules Verne was quick to capitalise on this and the main character in his book *The Adventures and Voyages of Captain Hatteras* (1866) '. . . under the influence of a magnetic force . . . was always walking towards the north'. The notion that birds, rather than people, might use a magnetic sense to navigate came from the Russian zoologist Alex von Middendorf in 1859 but he was given short shrift by most other ornithologists, including Britain's Alfred Newton, in the late 1800s.[9]

In 1936 Arthur Landsborough Thomson, another British ornithologist, wrote: 'No evidence of any magnetic sense has ever been obtained . . . moreover the suggestion becomes less attractive on examination, because the phenomena seem quite inadequate for the purpose.'[10] Similarly, in an otherwise insightful review, in 1944 Don Griffin said: 'no sensitivity to a magnetic field has been demonstrated in any animal, and sensitivity to as weak a field as the earth's is made extremely unlikely by the fact that living tissues are not known to contain any ferromagnetic substances (such as metallic iron oxide . . .) which alone are capable of exerting appreciable mechanical forces in the earth's magnetic field'.[II]

Not long after this, in the early 1950s, the German ornithologist Gustav Kramer started thinking about the problem in a new way, realising that navigation comprises two steps. Birds have to know where they are at the point of release, and they also have to know the direction of 'home'. This is how humans orientate: the first step involves studying a map (where am I?), the second involves using a compass (which way is home?). This became known as Kramer's 'map and compass' model.

There are several potential compasses. The one we are most familiar with is the magnetic compass, an instrument whose magnetised needle aligns itself with the field lines, or lines of force of the earth's magnetic field, and points north. Migration biologists have identified other types of compass that birds use to navigate, including a sun compass – used by birds that migrate during the day – and a star compass – used by nocturnal migrants.

The first evidence that birds might possess a magnetic compass emerged in the 1950s while Frederick Merkel and his student Wolfgang Wiltschko were studying the migration behaviour of European robins in Germany. Obviously, observing the process of migration can be difficult, especially with birds like robins which migrate at night. However, by capturing robins just before they set off on migration and placing them for a few hours in a specially designed 'orientation cage' – a precursor of the Emlen funnel – researchers could see in which direction they hopped or fluttered, behaviour that perfectly reflects their migration direction. Using orientation cages from which the robins could see the night sky, Merkel and Wiltschko found that the birds use the stars as their compass to maintain a south-westerly heading from Germany during their autumnal migration. However, when they looked at what robins did in *complete darkness*, they found that, far from being disoriented, which is what they expected, the birds continued to hop in their usual south-westerly direction. The implications were stunning: the stars were not essential for the birds to orientate themselves correctly. There had to be something else.

To test whether this 'something else' was a magnetic compass, they placed robins in orientation cages surrounded by a huge electro-magnetic coil, which allowed the researchers to alter the orientation of the magnetic field. They then compared the direction of the robins' hopping when the field was reversed or shifted to the east or west. As they hoped, the birds behaved exactly as if they were able to detect the magnetic field and altered the direction of their hopping accordingly.[12]

Studies of other species subsequently produced similar results and by the 1980s it was generally agreed – despite earlier scepticism – that birds do indeed possess a magnetic sense that allows them to read compass directions from the earth's magnetic field. In other words, these birds *do* possess a magnetic compass.

Remarkably, birds also possess a magnetic *map* that allows them to identify their location – like a GPS system, but, rather than using satellite signals, birds use the earth's magnetic field.[13] Migratory birds are not unique in this respect: a magnetic sense has been detected in non-migratory birds like the chicken, as well as in mammals and butterflies, presumably to help them navigate over more modest distances.[14]

One reason why a magnetic sense had once seemed so improbable was that birds do not obviously possess a specific organ capable of detecting a magnetic field. For senses such as vision and hearing, the eye and the ear are very obviously designed to detect light and sound, respectively, directly from the environment. Magnetic sensations are different because, unlike light and sound, they can pass through body tissues. This means that it is possible for a bird (or other organism) to detect magnetic fields via chemical reactions inside individual cells throughout its entire body.

There are currently three main theories as to how animals, including birds, detect magnetic fields. The first is referred to as 'electromagnetic induction' and may occur in fish, but birds and other animals seem to lack the highly sensitive receptors necessary for this mechanism. The second involves the magnetic mineral known as magnetite (a form of iron oxide), discovered in certain bacteria in the 1970s, which is responsible for bacteria aligning themselves with a magnetic field. Further research revealed that other species, including honeybees, fish and birds, also possessed minute crystals of magnetite. Microscopic crystals of magnetite were detected around the eye and in the nasal cavity of the upper beak – the latter inside nerve endings – of pigeons in the 1980s. As we'll see, these were promising locations if the crystals were to play a part in navigation.[15] The third theory comprises the interesting possibility that a magnetic sense might be mediated by a chemical reaction.

In the 1970s it was discovered that certain types of chemical reaction could be modified by magnetic fields, but at the time no one imagined that such a process might help migrating birds find their way. Even more remarkably, it seems that these particular chemical reactions are induced by light, prompting a group of researchers in the United States to speculate that birds might be able 'see' the earth's magnetic field.[16]

This unlikely idea prompted Wolfgang Wiltschko and his wife, Roswitha, to investigate. They knew from the research of others that free-flying pigeons with one eye covered by an opaque patch were better at homing if they could see with their right eye, rather than with their left. Significantly, this better right-eye- performance was most pronounced under cloudy conditions (when the sun was not visible). This meant, of course, that the birds could not be using a sun compass, but suggested that perhaps they were using a magnetic sense somehow linked to the right eye. It sounds unlikely, but the Wiltschko team also knew that birds' brains are highly lateralised, and the pigeon result was consistent with the left brain (which receives visual information from the right eye, as we saw in chapter 1) being better at processing information relating to homing and navigation. To test this idea directly, the Wiltschkos turned to their favourite study species once again, the European robin.

With both eyes uncovered, the robins hopped in their normal migratory direction. But when the magnetic field was experimentally switched through 180° (as in their earlier experiments), the birds also switched their hopping direction by 180°. The robins were then tested with one eye covered by an opaque patch. With the right eye exposed to light (i.e. a patch over the left eye), the birds' orientation was exactly as it had been with both eyes receiving light. But with the right eye covered and only the left eye receiving light, the robins were unable to orientate – implying that they could not detect the earth's magnetic field. This extraordinary result suggested that only the right eye can sense the earth's magnetic field.

How does this right-eye/left-brain process work? Is it simply that the right eye is more sensitive to light? To find out, the Wiltschkos conducted a further test, putting the equivalent of contact lenses on their robins. Both 'lenses' allowed the same amount of light into the eye, but one lens was frosted, giving a fuzzy image of the world, while the other was clear. The results were once again startling. The right-eye/left-brain effect remained, but when the robins viewed the world only through a frosted lens over the right eye they were unable to orientate. With a clear lens on the right eye, they orientated with precision, as before.

What this means is that light itself is not crucial, but what matters is the clarity of the image. It appears that it is the robin's ability to

see contours and edges in the landscape that provides the appropriate signal to trigger the magnetic sense. Extraordinary! As one of my
colleagues said: 'You couldn't make this stuff up.'

Where does this visually induced chemical reaction leave the
magnetite idea that I mentioned earlier? They do not seem to be
alternatives, but, rather, two separate processes that might work in
unison in the same animal: the chemical mechanism based in the
eye provides the *compass*, while the magnetite receptors in the beak
provide the *map*. The compass may detect the *direction* of the
magnetic field while the map detects the *strength* of the magnetic
field, and by integrating both types of information birds can find
their way home, whether it is across a featureless ocean or crossing
large land masses.[17]

The fact that a magnetic sense in birds was once considered
impossible, and that discoveries about the senses of birds are still
being made, is extraordinary. It is discoveries like this that make
science buzz.

7

Emotions

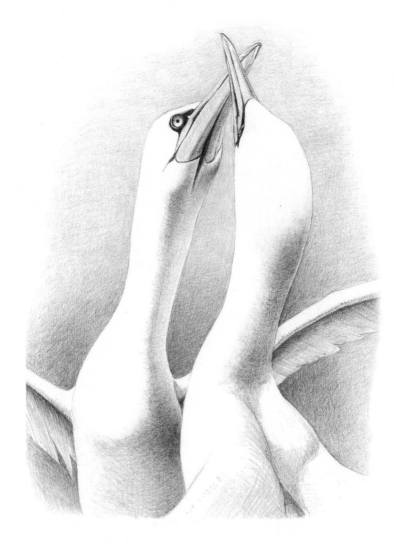

The greeting display of a pair of northern gannets – what do partners feel on being reunited?

*Many scientists appear to be uncomfortable about using the term emotion
when referring to animals, for fear that they automatically imply anthropomorphic assumptions of human-like subjective experience.*
Paul, Harding and Mendl, 2005, 'Measuring emotional processes in
animals: the utility of a cognitive approach', *Neuroscience and
Behavioral Reviews*, 29: 469–91

Resolute, on Cornwallis Island in Canada's Nunavut, is one of
the most remote settlements in the world. Almost everyone
who conducts research in the Canadian High Arctic arrives here
first by jet, then takes a light plane or helicopter to their final desti-
nation. As the jet descends, I see on either side of the runway the
remains of aircraft whose landing or take-off failed. This is my
stressful introduction to the Arctic. But worse awaits. Far from
fulfilling my romantic notion of the far north, I'm disappointed by
the desolate, muddy landscape, by the all-pervading smell of avia-
tion fuel and, most of all, by the casual way the local Inuit use birds
for target practice.

My arrival in mid-June coincides with the spring thaw and on
that first day I notice a pair of brent geese by a frozen pool: black
silhouettes against an icy background, waiting for the snow to melt
and the opportunity to breed. The next day I drive past the frozen
pool again, but am saddened to see that one of the geese has been
shot. Beside its lifeless form stands the bird's partner. A week later I
pass the same pond again, and the two birds, one live and one dead,
are still there. I left Resolute that day so I'm afraid I don't know
how long the bird stood vigil over its dead partner.

Is the bond that kept those two individuals together in life – and
in death – an emotional one, or is it simply an automatic response

that programmes birds like geese to remain close to their partner at all times?

Charles Darwin was in no doubt that animals like birds and mammals experienced emotions. In *The Expression of the Emotions in Man and Animals* (1872) he recognises six universal emotions – fear, anger, disgust, surprise, sadness and happiness – to which others later added jealousy, sympathy, guilt, pride and so on. Effectively, Darwin envisaged a continuum of emotions from pleasure to displeasure. Most of Darwin's book is about humans, and his own children in particular, whose facial expressions he studied in detail. Darwin also gained tremendous insights from his pet dog – and dogs, as every owner knows, make their feelings very obvious.

Like some of his predecessors, Darwin considered the vocalisations of birds an expression of their emotions. The sounds birds make under different circumstances have a quality that we identify with – harsh when aggressive, soft when directed to a partner, or plaintive when grabbed by a predator – making it easy for us to be anthropomorphic. In a similar vein, because we find birdsong enjoyable, it was long assumed that birds felt the same way and therefore sang for pleasure, either their own or their partner's.[1] At one level this is completely anthropomorphic. On the other hand, because we share both some ancestry and many sensory modalities with birds, it is just possible that we share a common emotionality.

Emotions often seem to run high when birds and their offspring interact. Parent birds care for their chicks, feeding them, allopreening them, removing their faeces and protecting them from predators. The injury-feigning display performed by ground-nesting birds like plovers and partridges provides a dramatic example of parental protection. Confronted by a fox or human, the parent bird drags one wing across the ground, creating the impression of injury and drawing the predator away from the more vulnerable chicks. Once considered to be a sign of parental devotion and intelligence, this distraction behaviour is now regarded as instinctive and assumed to be devoid of any emotional input, motivated simply by the conflicting tendencies of the need to stay close to the offspring and to escape from the predator.[2]

Even so, the way that parent birds protect their young, or the way chicks and ducklings follow their mother and run to her in times of danger, often makes it look as though they are tied together by an emotional bond. There is certainly a bond, but whether it is an emotional one is less clear. The bond is largely the result of the young bird imprinting on its mother soon after it emerges from the shell. Yet when chicks are hatched in an incubator, they will imprint on *anything* they first see, including inanimate objects such as a boot or a football. When this occurs we interpret the behaviour completely differently and ask ourselves how a young bird could be so stupid – how could it be emotionally tied to a boot or a ball? Yet there is an entirely logical explanation for this apparently 'stupid behaviour'.

Natural selection has favoured chicks that imprint on the first thing they see, for normally that is the mother, and under normal circumstances that works perfectly well. Rearing a chick with a boot or a ball, we are merely exploiting a simple inbuilt rule: follow the first thing you see. A cuckoo chick exploits the care of its foster parents in exactly the same way, exploiting their rule to feed anything that begs inside their nest. We might just as easily ask how the foster parents could be so stupid as to be duped by the young cuckoo.

It is clearly possible to account for parental and other behaviours without imputing emotions, but how certain can we be that birds and other animals don't experience emotions as we do?

Before considering the issue of whether birds experience emotions, I need to give you a bit of background. We'll start in the 1930s which, despite the promising start made by Darwin, is when the study of animal behaviour really began to take off. Researchers in North America adopted a hard-nosed psychological approach to behaviour, focusing mainly on captive animals and training them to tap at keys for rewards or to avoid punishment. For the 'behaviourists', as these researchers became known, animals were little more than automata. This is paradoxical, for the behaviourists' rationale relied on animals being able to respond to pain and to appreciate rewards. Most of today's animal behaviour students view the behaviourists' approach with disdain because it seems so artificial, but it did reveal a great deal about the cognitive capacity of animals. Pigeons, for example, were found to be at least as good as humans at memorising and categorising visual images. At the time, this seemed bizarre because pigeons appeared to be so inept at other tests, but when it was later realised that pigeons rely on visual maps to navigate, as we have seen, it made perfect sense.

The Europeans adopted a more naturalistic approach to behaviour, studying animals in their natural environment, creating the discipline of 'ethology'. Their initial focus was on what *caused* behaviour – what is the stimulus that triggers a behavioural response? A famous example from that era is the way a herring gull chick pecks at the red spot on its parent's beak, stimulating it to regurgitate food. Essentially, the ethologists were studying communication – what are animals saying to each other and what motivates them to behave in certain ways?

Even though the ethologists' approach was more naturalistic, it was also objective for they were desperate to avoid the trap of anthropomorphism, as Niko Tinbergen, one of ethology's main architects, explains in the introduction to *The Study of Instinct* (1951):

> Knowing that humans often experience intense emotions during certain phases of behaviour, and noticing that the behaviour of many animals often resembles our 'emotional' behaviour, they conclude that animals experience emotions

similar to our own. Many go even farther and maintain that emotions . . . are causal factors in the scientific meaning of the word . . . This is not the method we shall follow in our study of animal behaviour.

This view persisted well into the 1980s when researchers were '. . . advised to study the behaviour rather than attempting to get at any underlying emotion'.[3]

Some, however, like the eminent biologist Don Griffin, whom we have met already, were confident enough to challenge this view. His book *The Question of Animal Awareness*, published in 1976, was the first seriously to address the issue of animal consciousness and understand the 'mind' behind the behaviour.[4] Griffin's book was greeted with widespread derision, partly, as one colleague said, 'because his critics continue to define consciousness in a way that excludes the possibility that we can find out if it exists in animals'.[5] Nonetheless, throughout the mid-1970s and into the 1980s, there was a groundswell of interest in animal awareness fuelled largely by increasing concern over the issue of sentience and welfare in non-humans.[6]

Emotions, feelings, awareness, sentience and consciousness are all difficult concepts. They are tricky to define in ourselves, so is it any wonder they are difficult in birds and other non-human animals? Consciousness is one of the big remaining questions in science, making it both an exciting but also a highly contentious area of research.[7] Defining exactly what we mean by 'consciousness' or by 'feelings' is problematical, but nothing compared with the difficulty of trying to imagine how the mere firing of neurons can create a sense of awareness, or feelings of discomfort or euphoria.

These difficulties have not stopped researchers from trying to understand the emotional life of birds and other animals, but the lack of a clear conceptual framework has resulted in something of a free-for-all. Some researchers, for example, believe that birds and mammals experience the same range of emotions as we do. Others are more conservative, arguing that only humans experience consciousness, so humans alone are capable of experiencing emotions. Controversy is a normal part of science, and it tends to be greatest when there's much at stake. Consciousness is a major challenge, which means that this is an exciting time to be trying to understand the kinds of feeling birds and other animals experience. In humans, consciousness integrates the different senses. I have no doubt that the senses of birds are integrated as well, and that this integration creates feelings (of some sort) that allow birds to go about their daily lives, but whether they create consciousness as we understand it remains unknown. We have made a lot of progress in the last twenty years and the more we find out, the more likely it seems that birds do have feelings. But this is difficult research: difficult, but potentially very rewarding, for by gaining a better understanding of birds, whose lives are similar in many ways to our own – in terms of being predominantly visual, basically monogamous and highly social – we stand to gain a better understanding of ourselves.

Biologists, psychologists and philosophers have argued over the issues of consciousness and feelings for years, so I cannot hope to resolve them here. Instead, I shall use a very simple approach that allows us to think about what might be going on in a bird's head. The approach is based on the idea that emotions have evolved from basic physiological mechanisms that, on the one hand, allow animals to avoid harm or pain and, on the other, allows them to obtain things that they need, a 'reward', such as a partner or food.[8] Imagining a continuum, with displeasure and pain at one end and pleasure and rewards at the other, provides a good starting point from which to look at emotions.

Anything that upsets an animal's normal equilibrium is likely to be stressful. To put it another way, stress is a symptom of thwarted emotions. Hunger is a primary feeling that motivates us to seek food and failure to obtain it, especially in the long term, results in stress. Avoiding predators is what many animals spend much of their life doing, and being chased by a predator is stressful. Birds respond to stress by releasing the hormone corticosterone from the adrenal glands (located at the 'head' end of the kidneys in the abdomen) which in turn triggers the release of glucose and fat into the bloodstream, and provides the bird with a pulse of extra energy to minimise the impact of the stressful event. The stress response is therefore an adaptive one – it is designed to promote survival. However, if the stress is persistent the response can become patho-logical, resulting in loss of body weight, down-regulation of the immune system, a general decline in health and a complete loss of interest in reproduction.

The guillemots that have played such an important role in my research breed at exceptionally high densities, and the close proxim-ity of neighbours is the key to their breeding success since it enables them to avoid attacks on their eggs and young from gulls and ravens. A phalanx of guillemot beaks can deter most predators, but to be effective the birds have to be tightly packed together. Guillemots breed at exactly the same tiny site just a few centimetres square, year after year – sometimes for twenty years or more. Not surprisingly, they get to know their immediate neighbours very well, and specific relationships develop – friendships, possibly – mediated by allopreening. Sometimes these friendships pay off in an unexpected way. Occasionally, as a greater black-backed gull attempts to take guillemot eggs or chicks, I have seen an individual guillemot rush from the back of the group to attack the gull. This is an extremely risky venture since these huge gulls are quite capable of killing adult guillemots.[9]

Guillemots also look out for each other's offspring in another way. If a parent guillemot leaves its chick unattended, a neighbour will usually brood the chick – keeping it warm and keeping it safe from predatory gulls.[10] This form of communal care is rare among seabirds. In most other species unattended chicks would simply be eaten.

For guillemots breeding on the Isle of May, on the east coast of Scotland in 2007, something extraordinary happened. The sand eels they rely on to feed themselves and their chicks were in very short supply, and there was nothing else. In hundreds of field seasons of guillemot-watching by dozens of researchers at many different colonies, nothing quite like this had been seen before. As the parent birds on the Isle of May struggled to find food for their starving chicks, their normally harmonious behaviour disintegrated into chaos. Many adult guillemots were forced to leave their chicks unattended as they searched further afield for food, but their neighbours, instead of sheltering and protecting the unattended chicks, attacked them. Kate Ashbrook, who was studying the guillemots there, told me this:

> I remember watching in horror as one chick, stumbling into a puddle to escape attacking adults, was repeatedly forced face-down into the muddy water by pecks from a different adult. After a couple of minutes the attacker gave up and the chick struggled to stand up, but it was too weak and died shortly afterwards. It became just one of the many muddy little bodies that littered the breeding ledges. Other chicks were picked up by neighbours and swung around in the air, before being tossed off the cliff. These attacks were shocking and extremely tragic.[11]

This unprecedented anti-social behaviour seems to have been a direct result of chronic stress caused by the severe lack of food. In the following years, the food situation improved, and these same individual adult guillemots resumed their normal amicable behaviour.[12]

A similar response to a shortage of food has been seen in another bird, the white-winged chough. John Gould, one of the first ornithologists in Australia, commented in the 1840s on this species' intense sociality: 'It is usually met with in small troops of from six to ten in number feeding upon the ground . . . the entire troop keeping together . . . and searching for food with the most scrutinising care.' Gould came close to recognising that the chough was what we now refer to as a co-operative breeder, a species in which a breeding pair is assisted by non-breeding individuals, called helpers.[13]

Groups of white-winged choughs, comprising between four and twenty individuals, often remain together for years. They consist of a breeding pair, and the offspring from several previous breeding seasons, and sometimes unrelated individuals as well. All group members help to construct the bizarre mud nest – unlike any made by a European bird, it is a robust cup glued on to a narrow horizontal branch some ten metres above the ground – and all individuals take turns to incubate and to feed the chicks. Co-operative breeding is rare in Europe and North America, but is common among Australian birds, and the white-winged chough is an extreme example in that it is *always* co-operative. The species simply cannot reproduce as a conventional pair. The explanation lies in the chough's habitat. Digging for worms or beetle larvae in the dry ground is hard work. Young choughs rely on their parents for food for eight months – eight times longer than almost any other bird. Even after the adults have ceased to feed them, young choughs take several years to perfect their foraging skills. Essentially they serve a foraging apprenticeship in their parents' territory, and in return they perform household duties – defending the territory, watching for predators and helping at the nest. Food is so difficult to acquire that a minimum of two helpers is needed for a breeding pair to have any chance of rearing offspring. When researchers provided choughs with additional food, breeding success soared – confirming that the difficulty of finding sufficient food really does limit the birds' activities.

Chough groups function because a suite of behaviours keeps individuals bonded together. The birds do everything together: playing, roosting, dust-bathing and, during periods of rest, the birds line up along a horizontal branch and allopreen each other. What has this got to do with emotions? Being part of a tightly-knit group hinges on social interactions, both with other group members but also with the individuals of other groups. As Rob Heinsohn, who studied these birds for twenty years, said: 'The chough's chronic need for help leads to fascinating politics, especially when the weather turns bad.'[14]

As drought kicks in, the choughs experience several things simultaneously. The shortage of food increases their stress levels; the birds are forced to spend more time searching for food, and less time keeping an eye open for predators. If food is really short the birds use up all their body fat and start to use the protein reserves in their breast muscles. This in turn impairs their ability to fly, so that if a predator such as a wedge-tailed eagle does attack they have less chance of escaping. Stress is increased further as birds squabble over food. Whereas group members might once have shared food, as hunger bites individuals become extremely selfish and try to keep food for themselves. Larger or more dominant birds simply push the smaller individuals aside and steal their food; resistance is useless, for the stress of losing a fight may be more damaging still.

The difficulty of finding enough food under drought conditions eventually causes groups to disintegrate. The social bonds that once held individuals together dissolve – presumably in a sea of stress hormones – and individuals break off into smaller units, scouring the dry countryside for food. While such a tactic may increase the chances of finding something to eat, it also renders individuals more vulnerable both to harassment from other choughs and to attacks by predators.

Like many other birds, choughs probably have an innate response to the sight of an aerial predator like an eagle or falcon, by giving an alarm call and taking evasive action. Ethologists in the 1930s and 1940s studying this behaviour in young chickens and geese worked out exactly what it was about a shape moving overhead that triggered the response: it was a long tail, short neck and long wings.[15] Later, in 2002, researchers showed that the sight of a predator flying overhead (actually a model) resulted in an increase in the stress hormone corticosterone in the bloodstream, suggesting that birds experience the sensation of fear.[16]

The value of using a hormonal measure of stress rather than relying simply on behaviour to infer what kind of emotion the birds were experiencing was shown in a clever experiment in which captive, wild-caught great tits were allowed to see, on separate occasions, a Tengmalm's owl (a serious predator of great tits and other small birds) and a brambling (a finch, and no threat to the great tit). The great tits' behavioural responses to the owl and the brambling were identical, but only the owl elicited a surge of the stress hormone corticosterone, clear evidence that the great tits were more frightened by the owl.[17]

The rise in corticosterone in response to stress occurs rapidly, but decreases slowly. Researchers investigating the stress response in birds have employed a simple and harmless assay, which comprises holding the bird in the hand. On being held, the bird's heart rate, breathing and corticosterone levels all increase, and it is assumed that the bird responds much as it would do if it were captured by a predator. In other words, all three physiological changes indicate that the bird is frightened. While the increase in heart rate and breathing occur within seconds, it takes about three minutes for corticosterone to appear in the blood. Similarly, after the bird is released, heart rate and breathing return to normal within a few minutes, but depending on how stressful the experience, it may take several hours for the corticosterone to return to its normal levels.

An increase in corticosterone is a general response to any kind of stress. In snakes – and I'm using a reptilian example here because there is no equivalent information for birds – males that lose a fight with another male (over a female) experience a surge in corticosterone, and as a result are much less interested in sex for several hours than are the winners.[18]

That birds experience a similar physiological change in response to losing an aggressive interaction is suggested by a study of captive great tits in which exposing birds to an especially aggressive male in a cage for a few moments resulted in an increase in body temperature and a reduction in activity that lasted twenty-four hours. Similar results have been obtained in laboratory rats. Such tests are, by necessity, artificial since, however dramatic the results seem to be, the study birds are unable to 'escape' as they would in the wild. So, while such studies tell us that birds and other animals experience 'fear', the chances are that in the wild such effects are much smaller and the animals recover much more rapidly than they do in captivity.[19]

Studying wild zebra finches in Australia I spent many hours sitting quietly in a hide watching the birds through binoculars or a telescope. Inevitably, I saw lots of other wildlife during those hours, including one spectacular predation event. Galahs – pink and grey parrots – were common in the study area and often flew in front of where I sat, squawking as they went. On one occasion a brown falcon dropped out of the sky and chased the galahs. The flock took evasive action, but the falcon quickly singled out one of the birds, grabbing it in mid-air in a puff of pink feathers. The seized parrot shrieked abominably and even after the two birds disappeared into the trees I could still hear the parrot's plaintive cries, leaving me no doubt that the parrot was both terrified and in pain. My views, however, were later modified by another predation event I witnessed.

A puffin stepped out of its burrow at exactly the moment that a female peregrine was gliding along the cliff top. The falcon simply landed on top of the puffin, grasping it in its yellow talons. I know from capturing puffins myself that they are feisty and possess a powerful beak and sharp claws, so for a moment I thought the puffin might be able to escape. It didn't. Instead, it lay still, looking up at its captor, which, avoiding its gaze, stared resolutely out to sea. I imagine that the peregrine was waiting for its powerful clenching claws to do their business and for the puffin to die. It didn't. Puffins are tough birds, built to withstand intense pressure while diving for their own prey, and capable of withstanding high seas and gale-force winds. It was a stalemate. Five minutes passed with no obvious resolution in sight. The peregrine continued to gaze out to sea. The puffin wriggled slightly, its eyes were bright and it still looked full of life. Watching through my telescope, it was like a traffic accident, simultaneously appalling and compelling. Eventually, after fifteen minutes, the falcon started to pluck the breast feathers from the puffin, and five minutes after that began to eat the muscle from the puffin's breast. Only after the peregrine had eaten its fill, a full thirty minutes after capture, did the puffin eventually expire. Did it feel any pain? I don't know, for at no point during this grisly spectacle did the puffin show any sign of distress.

Jeremy Bentham (1748–1832), an early advocate of animal welfare, is perhaps best known for pointing out that the question is not whether animals can reason, but whether they can suffer.[20] It was, and remains, an important point, and Bentham was motivated by the fact that slaves were often treated appallingly and often no better than animals. A century earlier it suited the philosopher René Descartes to assume that animals were incapable of suffering, since denying the existence of pain helped to distinguish animals from ourselves, something the Catholic Church was keen to do. It also meant that animals could be abused without guilt. For others, like the naturalist John Ray, Descartes' contemporary, it was unimaginable that animals were without feelings. Why else did dogs cry during vivisection, he asked? The evidence seems irrefutable, yet objectively demonstrating the existence of pain in animals such as birds is tricky.[21]

Some researchers think that birds are capable of feeling only a certain type of pain. Imagine you have inadvertently placed your hand on the hotplate of a cooker. Your first reaction is a sharp sensation of pain followed by the immediate withdrawal of your hand. This is an *un*conscious reflex. It works via the pain receptors, the so-called nociceptors ('noci' refers to injury) in the skin, sending a signal to the spinal cord that triggers the reflex that results in your removing your hand. This is the first 'level' of the pain response. The second is the transmission of a message between your hand via your nerves to your brain where information is processed to create the sensation or feeling of pain. This is conscious pain – what you feel *after* removing your hand from the hotplate. It has been suggested that to feel this kind of pain requires consciousness. If, as some researchers propose, birds do not have consciousness, they cannot experience this particular 'feeling' of pain.[22]

This view presupposes that the unconscious pain reflex alone is sufficient for survival. Indeed, many other animals – both vertebrates and invertebrates – show the same kind of withdrawal reflex to unpleasant stimuli.[23] In terms of self-preservation the value of such a reflex is obvious. One has only to think of those unfortunate people who, because of a genetic mutation, are incapable of feeling pain and who routinely bite their tongue and cheeks while eating, or the Pakistani boy who makes a 'living' out of his inability to feel pain, by sticking knives into his arm for money.[24]

Studies of chickens, however, provide rather convincing evidence that birds can experience the *feeling* of pain. Chickens kept commercially at high densities often resort to pecking each others' feathers, and sometimes to cannibalism. In an effort to prevent this, the poultry industry amputates the tips of the birds' beaks. On the basis of our earlier discussion of the sense of touch, you may be able to anticipate some of what's coming next.

Beak trimming is a rapid procedure performed with a heated blade that simultaneously cuts and cauterises the beak. It appears that trimming results in a period of initial pain lasting two to forty-eight seconds, followed by a pain-free period of several hours, which is then followed by a second, more prolonged, period of pain. This is similar to what we experience following a burn injury. The initial period of pain in chickens was demonstrated by measuring the discharge from two types of nerve fibres, simply referred to as A and C fibres, from the pain receptors. The A fibres are responsible for the rapid, reflex-like pain response; the C fibres for the later, longer-lasting pain sensation. Young fowl seem to experience less pain and to recover more rapidly from beak trimming than adult fowl. The older birds also seemed to experience more discomfort and fifty-six weeks after the operation they still avoided using their beaks, preening less and performing less exploratory pecking than birds whose beaks had not been trimmed.[25]

The important point here is that, apart from some head-shaking immediately after the operation which presumably reflects the initial period of pain, the birds did not show any *obvious* outward signs of discomfort. Only by measuring subtle differences in their behaviour and physiology was it possible to demonstrate the longer-lasting *feeling* of pain.

On a more positive note, I am sometimes asked what my favourite bird is. For a long time I felt that question to be futile, but my experience with one species in 2009 changed the way I think about this. If I was asked now, I would have no hesitation in saying the sylph hummingbird, a South American beauty. In fact, there are two sylph species, the long-tailed and the violet-tailed. As their name indicates, these are tiny, elegant hummers of the most exquisite proportions and the most extraordinary colour: iridescent metallic green on the crown, and, depending on which species, metallic green or blue under the chin, and brilliant cobalt blue or violet along the entire length of their elongated tails.

Encountering a long-tailed sylph in Ecuador for the first time gave me the most extraordinary buzz, which lasted several days. The sylph was so exquisite that I wanted to possess it, to capture and hold on to its beauty. A photograph isn't enough because it cannot do the bird justice, but also because a single image isn't adequate to capture the full essence of the bird. I understand now why Victorians wanted to fill cabinets with the still sparkling but lifeless bodies of hummingbirds – it goes some way to providing the multiple images necessary to make these birds so vibrantly captivating.

For an ardent birder, seeing a rare or beautiful bird is a little bit like falling in love. When people say they love birds it is because, on seeing a particular bird, they get a special buzz in their brain.

Love was once believed to be impervious to scientific investigation, but recent technological advances mean that neurobiologists now feel that they have a window through which they can view human love. Using fMRI scanning technology, researchers can literally see into the brain while a subject says what emotions he or she is experiencing. When someone in a scanner looks at a photograph of a person they are passionately in love with, very specific parts of their brain 'light up'. These are areas of increased blood flow, and hence increased brain activity, and lie in the cerebral cortex and in the subcortical regions which together are referred to as the 'emotional brain'. Significantly, they are also known to be part of the 'reward system' of the brain. Looking at a photograph of a much-loved partner or lover, the hypothalamus region of the brain releases substances known collectively as neuro-hormones, which, by providing a link between the nervous system and the endocrine system, stimulates the reward centres.[26] These neurohormones therefore play a vital role in the formation of rela-tionships. There are other effects when people fall in love: another neurohormone, called serotonin, falls to levels similar to those found in people suffering from obsessive-compulsive disorders, which may explain why lovers sometimes become single-minded and obsessive. Two other neurohormones, oxytocin and vasopres-sin, also produced by the hypothalamus (especially during orgasm), increase on falling in love, and also seem to play a vital role in bonding.

These findings had their origin not in a bird, but in a mammal, the prairie vole, one of just a handful of mammal species with a long-term pair bond and with shared paternal care. During sex, oxytocin and vasopressin are released from the vole's brain, facilitating and reinforcing their pair bond: oxytocin in the female, vasopressin in the male. If, however, the secretion of these two chemicals is experi-mentally blocked, the voles fail to form a relationship. Conversely, even without copulation, an injection of the two chemicals will result in a relationship being formed. Even more remarkably, when researchers introduced a gene that stimulates vasopressin secretion into males of a different, non-monogamous vole species, the meadow vole, it showed a significant increase in the tendency to form a pair bond with a female, suggesting that bonding may depend on a single gene. The researchers who conducted this work are keen to emphasise its preliminary nature, and that we should be careful of extrapolating to other species, but their results do suggest a mechanism linking pair bonding behaviour and the reward system in the brain.[27]

We do not yet know whether similar processes occur in birds. Currently, there are two research groups investigating this, both using the monogamous zebra finch as their study species. Although they have detected neurohormone activity in the appropriate parts of the brain, so far it is not clear that the same processes that occur in the prairie vole also occur in the zebra finch. The research is ongoing so we should soon know.[28]

The reward system is central to everything we as humans do. It is what keeps us going: it is why we eat, why we have sex and why some of us watch birds. The greatest pleasures that (most) humans can experience, however, are the emotional experiences associated with love and lust. Love can be both romantic and parental, and both forms involve 'attachment' or bonds: between partners, and between parents and offspring. Romantic love, of course, usually leads to physical desire and lust. It is easy to propose an adaptive explanation for love: a pair of individuals working together is more effective and successful in rearing offspring than individuals with any other breeding system – at least under certain ecological circumstances.[29]

Birds, too, are famously monogamous, by which I mean they are unusual among animals by breeding as pairs – a male and a female joining forces to rear offspring together. In a survey conducted in the 1960s, David Lack estimated that over 90 per cent of the 10,000 known species of birds reproduce in this way. The rest are either polygamous (a breeding system subdivided into *polygyny* comprising one male with several females, and, more rarely, *polyandry*, comprising one female and several males), or promiscuous, with no kind of bond whatsoever between male and female. Later, the notion of almost ubiquitous monogamy among birds had to be modified, when molecular paternity studies revealed just how

widespread extra-pair paternity is. Even though Lack was right
about the majority of birds breeding as pairs, monogamy does not
imply an exclusive sexual relationship. Copulations outside the pair
bond and extra-pair offspring are common, and ornithologists now
distinguish between what they call social monogamy (breeding as a
pair) and sexual monogamy. The latter is an exclusive mating
arrangement in which there is no infidelity, and is exemplified by
the mute swan and a relatively small number of other species.[30]

I am not going to speculate about the emotions that might be
involved in avian infidelity. However, it is worth thinking about the
emotions associated with pair bonds, especially the enduring bonds
of long-lived birds, and the bonds that occur among different
members of co-operatively breeding groups like white-winged
choughs, bee-eaters and long-tailed tits. In all cases there is likely to
be an emotional dimension to bonding. The problem is that, so far
at least, we have no way of unambiguously demonstrating such an
effect.[31]

Here is how it might work. There are several things that birds do
that we know are tightly associated with social relationships, both
with a partner, and, in co-operatively breeding species, with other
group members as well. These include greeting ceremonies, certain
vocal displays and, as we've seen, allopreening.

Whether the goose whose partner was shot near Resolute in
northern Canada experienced any emotional response to its loss is
something we do not know. Geese are normally long-lived with
long-term pair bonds and strong family ties – the young remain
with the parents for several months and the family even migrates
together. When pair members are temporarily separated, they typically perform a greeting display or 'ceremony' on being reunited.
Such displays are widespread among long-lived birds and are particularly protracted when pair members are reunited after a winter's
separation, in birds such as penguins, gannets and guillemots.

Throughout the breeding season, pair members greet each other, even after a relatively short absence, when one bird returns after a foraging trip. Strikingly, the duration and intensity of these greeting displays is closely tied to the length of time the pair members have been apart.[32]

Bryan Nelson, who has studied gannets and boobies throughout his career, described the North Atlantic gannet's meeting ceremony as 'one of the finest displays in the bird world'. If you visit a gannet colony, such as the Bass Rock in Scotland, you can see this display very easily. As one member of the pair returns to its partner at the nest, the two birds stand upright, breast to breast with outstretched wings, and skyward-pointing beaks. In a frenzy of excitement they clash their bills together, intermittently sweeping their head down over the neck of their partner, calling raucously all the time.

Under normal circumstances this greeting display lasts a minute or two, but Sarah Wanless, who studied gannets at Bempton Cliffs in northern England, observed a particularly prolonged instance. At one of the nests she was regularly checking, the female of the pair disappeared, leaving the male to care for the tiny chick alone, which, against all the odds, he did. One evening, the female returned after a remarkable five-week absence and luckily Sarah was there to witness it. To her amazement, the two birds performed an intense greeting ceremony that lasted a full seventeen minutes! Because the greeting ceremonies of humans (kissing, hugging, etc.) are also more elaborate the longer participants have been apart, it is tempting to assume that birds experience similar pleasurable emotions on being reunited.[33]

In many species, like the Eurasian bullfinch, pair members keep in contact while they are foraging in dense vegetation by their constant piping contact calls. In other species, including African shrikes, robin-chats, and certain tropical wrens, pair members perform antiphonal singing – an alternating duet so beautifully synchronised that it sounds like a single bird. The function of such duetting is not fully understood, but may serve in territory defence.[34] One of the most remarkable of all such displays is the 'carolling' by Australian magpies, which, like the white-winged chough, is a co-operative breeder. Carolling consists of the entire magpie group – some six to eight birds – standing on the ground, often around a bush or a fence post, and together uttering their hauntingly melodic song (which fans of the TV series *Neighbours* will be familiar with since it is often on the sound track). As Ellie Brown, who studied magpie carolling, said: 'Communal songs, like motets and madrigals, are made up of the combined melodies of all the singers.' In terms of function, Ellie likened it to a human war chant, creating and reinforcing the group cohesion necessary to maintain and defend their territory.[35]

Most co-operative breeders, many seabirds and small finches like the zebra finch spend a remarkable amount of time allopreening. In primates, the equivalent behaviour, allogrooming, is known to result in the release of endorphins, which result in the groomed individual appearing to be relaxed – presumably a pleasurable feeling.[36] The tame African grey parrots studied by Irene Pepperberg also seemed to go into a state that resembled 'relaxation', with half-closed eyes and a relaxed body posture, while she tickled or allopreened them. If she stopped, they would request 'tickle', but if she inadvertently touched a growing pinfeather, which is presumably very sensitive, they would give her a threat bite, then relax again and request 'tickle' again. Another parrot tamed and trained to talk by the French psychologist Michel Cabanac used the word '*bon*', good, in response to pleasurable events, including being preened or tickled – despite not being trained to do so.[37]

Our best hope for better understanding the kinds of feeling birds might experience is through a combination of careful behavioural studies, like those that looked at how much debeaked hens used their beaks, and physiological studies that measure response to what are likely to be emotional situations, such as greeting displays, allo-preening and separation from partners. Physiological measures include changes in heart rate, breathing rate, the release neuro-hormones from the bird's brain, or changes in brain activity visualised by scanning technology. None of this is easy, and at the present time cannot be done on free-living birds. Yet I can imagine that in the not-too-distant future it will be possible to measure at least some of these responses in wild birds. I will leave you to decide whether, on the basis of the science I have described here, birds experience emotions. My impression is that they do, but as Thomas Nagel said when he asked what it's like to be a bat, we can probably never know if birds experience emotions in the same way as we do.

Postscript

In this book I have discussed the different senses of birds one at a time. I have done this for convenience and clarity, but in reality, of course, the senses are used in combination. Psychologists have shown that we utilise and process information from several different sense organs simultaneously and often subconsciously. When we meet someone for the first time, for example, our primary source of information is visual, but almost without knowing it we assess how they smell, how they sound, and, if we embrace them or shake hands, how they feel, too (how I hate a limp handshake). It makes sense that birds must also integrate information from their different sense organs, because doing so provides them with more information that in turn may affect their survival.

Sometimes it can be hard for researchers to figure out exactly which senses birds are using to assess their environment. The sight of a thrush, blackbird or American robin foraging for earthworms on a suburban lawn is a familiar one. The bird hops forward, stops, cocks its head to one side and waits – is it looking or listening? Then, with a rapid lunge, it snatches a worm from the ground. In the 1960s the American ornithologist Frank Heppner studied the question of which sense American robins use to capture prey. He found that if he played 'white noise' to captive robins while they were foraging for worms, it made absolutely no difference to their foraging success. He concluded that robins hunted visually and that, when a bird cocked its head, it was *looking* rather than *listening*, using one eye to scan the ground for signs of a worm.[1]

Thirty years later, Bob Montgomerie and Pat Weatherhead revisited this problem and came to rather different conclusions. They agreed that the head-cocked posture was entirely consistent with looking and that the angle of the bird's head meant that the image of the ground was projected directly on to the bird's fovea. But when they removed all visual cues – holes in the ground or earthworm casts – the birds were still able to find their prey. By a process of elimination, Montgomerie and Weatherhead showed that the robins found food by *hearing* the worms. If you put your ear over an earthworm's burrow you can sometimes hear the worm's tiny bristles rustling against the sides.

They also discovered that Heppner's study was flawed because the birds could actually see the worms in their holes, so it was hardly a case of discovering how the birds detected 'invisible' prey. The take-home message from Montgomerie and Weatherhead's study is an important one. It is this: even though our interpretation of a particular behaviour suggests that birds are using one particular sense it requires careful experimentation to be absolutely sure which one it is.[2] Outside the laboratory American robins undoubtedly make use of both vision *and* hearing while hunting. They may also use smell; they may even detect the worm's movements in the soil through touch sensors in their legs and feet.

More spectacular than the American robin's ability to detect worms is the facility with which arid-region waterbirds can sense rain falling hundreds of kilometres away. Thousands of greater and lesser flamingos suddenly appear within hours of rain falling at Etosha Pan, Namibia or the Makgadikgadi Pans in Botswana. In these arid regions rain is erratic, but once it falls the shallow pans fill rapidly with water. Those flamingos spend the winter at the coast, and without directly experiencing any rain themselves are somehow able to tell that rain has fallen and in response fly inland. They can not only detect distant rain, but also appear to be able to tell *how much* rain has fallen, abandoning their coastal winter quarters only if the rainfall is sufficient for breeding. Are the flamingos responding to the vibration of distant thunder? Possibly, but they often respond to distant rain even when there has been no thunder. Are they responding to the sight of towering cumulus rain clouds, visible from considerable distances on the ground and further still from the air? Are they responding to changes in barometric pressure?[3]

So far, no one knows what senses flamingos and other birds use to detect distant rain. Stephen Jay Gould's essay 'The Flamingo's Smile' celebrates the fact that flamingos feed with their heads upside down, filtering tiny prey items from the water. Gould assumed that the flamingo's enigmatic smile was a consequence of its upside-down bill, but I prefer to think that they are amused by our puzzlement over their mysterious ability to sense distant rains.[4]

The clearest example of the way our own senses are used in combination relates to taste. If you hold your nose (and therefore temporarily remove your sense of smell) and bite into a (peeled) onion, you can do so without tasting it. Stop holding your nose and the taste of the onion instantly becomes apparent. Psychologists reckon that 80 per cent of taste occurs via our sense of smell. Taste and vision are also intimately linked, and brain scans show that simply looking at food lights up the taste regions of the brain. Do similar interactions occur in birds' brains? The experiments are more difficult to do, of course, but it would be interesting to know.

The other well-known feature of the human sensory system is 'compensatory enhancement' (or, more technically, cross-modal plasticity) – the ability to develop certain senses if one is impaired or lost. There are two explanations for this. One is that without the ability to see, for example, people simply pay more attention to sounds or other sensory inputs. The other is that, deprived of one sense, the brain reorganises itself to enhance the other senses. Both seem to be true. The fact that the brain can reorganise itself in this way is compelling evidence for the sophisticated integration of sensory information. I wondered whether the ability of our blind zebra finch, Billie, to distinguish footsteps (see page 75) might be an example of this type of compensation, or whether a fully sighted zebra finch could do the same. It would have been relatively easy to check, but by the time I thought of it Billie had passed away.

One of the most impressive examples of compensatory enhancement is the ability of blind people to echolocate. Unsighted people often learn to navigate around their home by listening to the echoes of sounds bouncing off the furniture – a phenomenon known as *passive* echolocation because it does not require the person to make any noise. As I worked on this book, I thought about passive echolocation, and noticed that I was sensitive to echoes, too. In fact, I discovered (not very usefully) that I could tell as soon as I opened one particularly noisy door where I work (and without being able to see) whether there was someone already in there. Once I had detected this ability, each time I visited this room I tried to predict on opening the door whether I was right: my success rate was about 85 per cent. Far more impressive, though, is the fact that some blind people use *active* echolocation to enable them to go mountain biking. As they ride they click their tongues about twice a second, and using the echoes they hear are able to stay on the track and avoid obstacles![5] I described earlier how oilbirds and swiftlets actively echolocate inside dark caves, but I wonder whether other cave-dwelling or nocturnal birds might also employ passive echolocation.

Using our own sensory system provides our only starting point for understanding how birds experience the world, and as long as we recognise that they have senses that we do not possess, and as long as we don't automatically assume that even the senses they share with us are identical, then we can begin to gain some understanding of their world.

The ability to visually recognise individuals provides a nice example. We are extraordinarily good at recognising faces: we know within a fraction of a second if we've seen a particular face before, and we have an extraordinary capacity to recognise someone we know. In the chapter on seeing, I described an incident that suggested to me that, on the basis of vision alone, guillemots can identify their partner in flight at a distance of several hundred metres. This seems extraordinary not because the eye of a guillemot is that different from our own, but because to the human eye most guillemots are utterly indistinguishable even at point-blank range. My example is a mere anecdote, but it is consistent with other observations that suggest that guillemots and, indeed, many other birds are very good at recognising individuals. The most obvious and well-established way that birds recognise other individuals is through their voice. We know this because hearing lends itself to elegant testing through so-called playback experiments in which birds are played recordings of calls and songs (which exclude all other cues) to see how they react. Hundreds of such experiments show unequivocally that voice and hearing are important ways for birds to recognise each other.

Working out whether birds use other senses to identify individuals is rather more difficult, but, again, anecdotal evidence suggests that they do. The peck order in chickens, for example, relies on birds being able to recognise each other by sight. My colleagues Tom Pizzari, Charlie Cornwallis and I inadvertently demonstrated this in an unexpected way. We were conducting experiments to establish how many sperm cockerels transfer to hens during mating. If we presented the same hen to the same cockerel every few minutes over an hour or so, the number of sperm showed a predictable decline with each successive mating. If, however, we swapped the female halfway through the experiment, the male's sperm count shot up. Since the cockerels always seemed to look at the hen before mating, visual recognition seems the most likely explanation. Other birds are known to be capable of recognising other individuals by sight. The turnstone has an individually distinct pattern of black and white plumage on its head and upper body, and by making models painted to resemble particular individuals Philip Whitfield confirmed that visual cues were crucial in individual recognition. In more sophisticated tests in the laboratory, pigeons can also recognise other pigeons they see on a video screen.[6]

This ability of birds to visually recognise particular individuals, and sometimes from a distance, seems all the more remarkable in the light of other observations and experiments. The facts that young herring gulls can be duped into responding to a two-dimensional cardboard cutout of an adult gull's head, or that a buffalo weaver is willing to copulate with a model female consisting of little more than a wire frame with wings, or a duckling can imprint on a human (or a boot) and behave towards it as though it was its mother, all suggest that some fundamental differences in perception exist between birds and ourselves. However, a moment's reflection should make us cautious of jumping to such a conclusion. With only a modicum of imagination, we can probably think of human equivalents to all three of those bird examples. Our ability to be duped by our sensory system is extraordinary: we are fooled by holograms, befuddled by optical illusions such as Necker's cube, Penrose's triangle or Escher's endless staircases, and, because of the way our brain is wired, we are incapable of seeing an upside-down human face objectively. Understanding why our senses are fooled by such tricks has provided extraordinary insight into the way our sensory systems function. The same kind of approach might increase our understanding of the way birds perceive the world – as far as I am aware, no one has yet used it, but I guess they soon will.[7]

A psychologist recently commented that this – the early twenty-first century – is the golden age of sensory research in humans.[8] I like to think that the golden age of sensory research in birds is still to come. I have tried to summarise what we currently know and also what we don't know about the senses of birds. Our understanding of the human sensory system is advancing in leaps and bounds, and, if history is anything to go by – and I think it is – then it is inevitable that what we discover about the senses of humans will allow us to make similar studies of birds. History also shows very clearly that what we discover about birds (and other animals), including their seasonal remodelling of the brain, or their regeneration of hair cells in the inner ear, have huge implications for humans, too. At the present time we have a good basic understanding of at least some of the senses of birds, but the best is yet to come.

Notes

PREFACE

1. Some blind people are able to use echolocation to navigate within a room, and some – as mentioned in the Postscript – can echolocate outside, by uttering a double click noise and listening to the echo (Griffin, 1958; Rosenblum, 2010).
2. The invention of the microscope is usually credited to a Dutch father and son team, Hans and Zacharias Jansen, spectacle makers in the 1590s and 1600s, although the ancient Chinese reputedly made low-power 'microscopes' using a lens and a tube of water (probably of quartz) (Ruestow, 1996); fMRI: Voss et al. (2007).
3. 'Swifts', a poem by Ted Hughes.
4. Corfield et al. (2008).
5. Tinbergen (1963); Krebs and Davies (1997).
6. Forstmeier and Birkhead (2004)
7. Swaddle et al. (2008).
8. Eaton and Lanyon (2003).
9. Hill and McGraw (2006).

1. SEEING

1. 'Shrike' (*OED*) means shriek, and may refer to the cry the bird made on seeing a falcon, when being used by falconers. Linnaeus called it *Lanius* (butcher, hence butcher bird) *excubitor* (sentinel). Some believe that 'sentinel' refers to its use by falconers, but others think that it refers to the

bird's habit of sitting out in the open when it is hunting: Schlegl and Wulverhorst (1844–53); the quote is from Harting (1883).

2. Harting (1883).

3. Harting (1883).

4. Wood and Fyfe (1943); Montgomerie and Birkhead (2009); Wood (1931): note that Casey Wood worked with J. R. Slonaker, one of the pioneers in the study of avian eyes.

5. Walls (1942).

6. Wood (1917): the loggerhead shrike is very closely related to the great grey shrike.

7. Ings (2007); Nilsson and Pelger (1994);

8. Rochon-Duvigneaud (1943); Buffon (1770, vol. 1). The idea that the vision of birds is 'better' than that of humans is simplistic, partly because different bird species differ in their eyesight, and, because vision is multi-faceted, some birds have good visual acuity, others good sensitivity.

9. Rennie (1835: 8).

10. Fox et al. (1976).

11. One possibility is that birds have something equivalent to the inbuilt face recognition system that humans have (see Rosenblum, 2010), and while to us all guillemots look the same, to a guillemot every other guillemot looks different. Another possibility is that, like us, birds can recognise each other from the pattern of movement.

12. Harvey's book has been translated by Whitteridge (1981: 107).

13. Howland et al. (2004); Burton (2008).

14. Wood and Fyfe (1943: 600).

15. Walls (1942). As is now clear, kiwis have traded vision for a suite of other senses (see chapters 2, 3 and 5).

16. Derham (1713).

17. Woodson (1961).

18. Martin (1990).

19. Newton (1896: 229).

20. Wood and Fyfe (1943: 60).

21. Perrault (1680).

22. Ray (1678).

23. Perrault (1676, cited and illustrated in Cole (1944)).

24. Newton (1896); Wood (1917).

25. Soemmerring – cited in Slonaker (1897).

26. Known also as the temporal and lateral; deep and shallow foveas.

27. Snyder and Miller (1978).

28. But see Tucker (2000) and Tucker et al. (2000). Whether binocular vision (both eyes viewing the same object simultaneously) results in depth perception (stereopsis) in birds is unclear (Martin and Orsorio, 2008).

29. Martin and Osorio (2008):

30. Gilliard (1962). Note that this was the Guianan species of the cock-of-the-rock.

31. Andersson (1994).

32. Cuthill (2006).

33. Ballentine and Hill (2003).

34. Martin (1990).

35. Martin (1990).

36. Nottebohm (1977); Rogers (2008).

37. Thomas More (1653) mentions parrots being mainly left-handed; see also Harris (1969) and Rogers (2004). Handedness in crossbills, first noted by Townson (1799, cited in Knox, 1983) is associated with their crossed bill, which is an adaptation for extracting seeds from pine cones. In the common crossbill about half the population is 'left-billed', in which the lower bill crosses to the left of the upper; the rest are 'right-billed' birds. As Knox (1983) says: 'Because of the way the bird holds the cone, most of the strain is taken by the foot on the opposite side to that to which the lower mandible [bill] crosses. Therefore a left-billed bird is "right-handed". Right-handed birds have a longer right leg and larger jaw muscles on the left hand side of

the skull, so the asymmetry is quite pronounced. The direction of bill crossing is determined as nestlings before the bill tips actually cross. Neither the cause of the direction of crossing nor its cognitive consequences are known. The Hawaii akepa (a small red honey-creeper) also possesses a (subtly) crossed bill and exhibits handedness' (Knox, 1983).

38. Rogers (2008).
39. Lesley Rogers, personal communication.
40. Rogers (1982).
41. Rogers (2008); see also Tucker (2000), Tucker et al. (2000).
42. Weir et al. (2004); see also Rogers et al. (2004).
43. Rogers (1982).
44. Rattenborg et al. (1999, 2000): it is worth noting, with scientific caution, that knowing whether a bird is truly asleep requires knowledge of its brain function since sleep is defined by particular patterns of electrical brain activity. It isn't possible to know whether or not a bird is asleep simply by seeing whether its eye(s) are open or not.
45. Rattenborg et al. (1999, 2000).
46. Lack (1956); Rattenborg et al. (2000).
47. Stetson et al. (2007). In fact, insects achieve this by extracting only the relevant information they need from the flow of images they receive, and it is possible that birds may do something similar.

2. HEARING

1. Newton (1896: 178).
2. Bray and Thurlow (1942); Dooling (2000).
3. Baldner's (1666 – see also Baldner (1973) facsimile) illustrated account of the birds of the Rhine inspired Willughby and Ray (Ray, 1678). Baldner was incorrect in thinking that the bittern's boom was uttered mainly by the female, but correct in saying that the head is held high while booming. Others thought the sound was created by the bittern

blowing into a reed. Writing about 'The Fen Country' during his journey through Britain, Daniel Defoe commented: 'Here we had the uncouth musick of the bittern, a bird formerly counted ominous and presaging, and who, as fame tells us, (but as I believe no body knows) thrusts its bill into a reed, and then gives the dull, heavy groan or sound, like a sigh, which it does so loud, that with a deep base, like the sound of a gun at a great distance, 'tis heard two or three miles, (say the people) but perhaps not quite so far' (Defoe 1724–7). The South American bellbirds *Procnias* spp. also have an extremely loud call.

An ell(e) – a forearm, literally – was a unit of measurement previously used by tailors: however, it differed between regions of Germany. A forearm is about 40 cm, so Baldner's five ells would make 200 cm (two metres) – unlikely for the bittern's esophagus but possible if he meant the entire gut. There's a note in Baldner (1666) that says that a Strasbourg ell is '2 foot long, but a foot is a little less than an English foot' which simply adds to the confusion.

4. Best (2005).
5. Henry (1903).
6. Merton et al. (1984).
7. Brumm (2009).
8. Cole (1944: 433).
9. Pumphrey (1948: 194).
10. Thorpe (1961); Marler and Slabbekoorn (2004).
11. Intriguingly, in the present context, the term 'pinna' means feather, although the link with the mammalian ear is unclear.
12. An interesting exception is the woodcock *Scolopax* spp., whose ear opening lies below, but well in front of, the eye, possibly because so much space is taken up by their huge eyes, and this is the only position possible.
13. The ear coverts have a shiny appearance because these feathers lack the normal barbules, the tiny hooks that hold the filaments of other feathers together.

14. Sade et al. (2008).

15. http://www.nzetc.org/tm/scholarly/tei-Bio23Tuato1-t1-body-d4.html

16. Cole (1944: 111) makes the same point in criticising the limitations of Hieronymus Fabricius's seventeenth-century account of the ear: 'Neither did it occur to him [Fabricius] that the pinna might be a new formation characteristic in mammals. Whilst therefore it would be legitimate to enquire into the causes of its disappearance in some mammals, there was no occasion to attempt to explain its absence in birds, reptiles and fishes, where it had never existed.'

17. Saunders et al. (2000), cited in Marler and Slabbekoorn (2004: 207).

18. Bob Dooling, personal communication.

19. Pumphrey (1948).

20. Walsh et al. (2009).

21. White (1789).

22. Dooling et al. (2000).

23. Lucas (2007).

24. Hultcrantz et al. (2006); Collins (2000).

25. Dooling et al. (2000).

26. Marler (1959).

27. Tryon (1943).

28. Mikkola (1983).

29. Konishi (1973): the facial disc improves sound gathering by about ten decibels.

30. Pumphrey (1948); Payne (1971); Konishi (1973).

31. Konishi (1973).

32. Konishi (1973).

33. Hulse et al. (1997).

34. Morton (1975).

35. Handford and Nottebohm (1976).

36. Hunter and Krebs (1979).

37. Slabbekoorn and Peet (2003); Brumm (2004); Mockford and Marshall (2009).

38. Naguib (1995).
39. Ansley (1954).
40. Vallet et al. (1997); Draganuoi et al. (2002).
41. Dijkgraaf (1960).
42. Griffin (1958).
43. Galambos (1942).
44. Some bat species can hear higher frequencies: the tiny Percival's trident bat, *Cloeotis percivali* (it weighs just four grams), can hear frequencies of 200 kHz (Fenton and Bell, 1981).
45. Griffin (1976).
46. Humboldt, quoted in Griffin (1958: 279).
47. Griffin (1958).
48. Griffin (1958: 289; see also Konishi and Knudsen, 1979) – Griffin must have made an error: the frequency is about two kilohertz.
49. Griffin (1958).
50. Konishi and Knudsen (1979).
51. Griffin (1958: 291).
52. Ripley, cited in Griffin (1958).
53. Novick (1959).
54. Pumphrey, in Thomson (1964: 358).

3. TOUCH

1. Billie may have heard my daughter's footsteps, but he might also have felt them. Birds have special vibration dectectors in their feet and legs (Schwartzkopff, 1949), and it may be these that allow birds to sense branches trembling or, worse, to 'anticipate' earthquakes.
2. Our most touch-sensitive regions are our fingertips, lips and, to a lesser extent, our genitalia.
3. There is not much published on the touch receptors in the bills of small birds, but Herman Berkhoudt (personal communication) tells me that he examined a zebra finch bill and found many

touch receptors including (forgive the names) Merkel cell receptors, double-column Merkel cell receptors and many Herbst corpuscles, all indicative of a very sensitive bill tip.

4. Goujon (1869) refers to these as Pacinian corpuscles and they were first discovered in human fingers by Abraham Vater in the 1740s, but were mistakenly assumed to have been discovered by Filipo Pacini in 1831 and named (by others) in his honour.

5. Berkhoudt (1980).

6. Goujon (1869).

7. Berkhoudt (1980).

8. The quote is originally from Nathan Cobb (1859–1932), founder of the study of nematodes.

9. Berkhoudt (1980).

10. The Royal Society seems to have lost Clayton's drawing; Nichola Court searched for them on my behalf, but failed to find them. William Paley (*Natural Theology*, 1802, pp. 128–9) later used Clayton's information – with an accompanying illustration of his own – as evidence of God's wisdom. Paley plagiarised Ray's *Wisdom of God* (1961) and William Derham's *Physico-Theology* (1713): Derham had quoted Clayton's writing and had probably seen his illustrations of the nerves in the duck's beak.

11. Berkhoudt (1980).

12. H. Berkhoudt, personal communication.

13. Krulis (1978); Wild (1990).

14. H. Berkhoudt, personal communication. 'Touch' is a multi-faceted concept, reflecting the different types of receptors. The simplest are free nerve endings which detect pain and changes in temperature; slightly more complex are Merkel's tactile cells (which detect pressure); followed by Grandry bodies, which consist of two to four tactile cells and detect movement (velocity); and the lamellated Herbst corpuscles (similar to Vater-Pacinian corpuscles in mammals), which are sensitive to acceleration.

15. Brooke (1985); M. P. Harris had never seen guillemot allopreening result in the removal a tick, and even the addition of false ticks didn't elicit allopreening (M. P. Harris, personal communication).

16. Radford (2008).

17. Stowe et al. (2008).

18. Senevirante and Jones (2008).

19. Carvell and Simmons (1990).

20. Thomson (1964).

21. Pfeffer (1952); Necker (1985). The receptors associated with filoplumes, together with the numerous other touch receptors in a bird's skin, are also incredibly important in keeping the plumage sleek when the bird is flying. Indeed, birds have more touch receptors in their skin than mammals; and flying birds have more receptors per unit area than flightless birds, suggesting that they play a crucial role in flight (Homberger and de Silva, 2000).

22. Senevirante and Jones (2010).

23. They are also able to detect prey by smell and taste (see chapters 4 and 5); see also Gerritsen et al. (1983).

24. Piersma (1998).

25. Parker (1891); see also Cunningham et al. (2010) and Martin et al. (2007).

26. Buller (1873: 362, 2nd edition).

27. These are: dunlin, *C. alpina*, western sandpiper, *C. mauri*, and least sandpiper, *C. minutilla*: Piersma et al. (1998).

28. McCurrich (1930: 238).

29. Coiter (1572).

30. Sir Thomas Browne (c. 1662), *The Birds of Norfolk* – see Sayle (1927).

31. Following Willughby and Ray (Ray, 1678), a succession of anatomists and naturalists dissected woodpeckers, fascinated by their unusual tongue. These include: Jacobaeus (1676), Perrault (1680), Borelli (1681), Mery (1709), Waller (1716) – all cited in Cole (1944).

32. Buffon (1780: vol. 7).

33. Villard and Cuisin (2004).
34. Fitzpatrick et al. (2005); Hill (2007). The other evidence would be DNA from a moulted feather.
35. Wilson (1804 – 14: vol. 2).
36. Audubon (1831–9).
37. Audubon (1831–9).
38. Martin Lister, cited in Ray (1678); Drent (1975).
39. Lea and Klandorf (2002).
40. Drent (1975); Jones (2008); and D. Jones, personal communcation.
41. Alvarez del Toro (1971).
42. Friedmann (1955); Claire Spottiswoode showed me honeyguide chicks killing bee-eater nestlings in her field site in Zambia.
43. Jenner (1788); Davies (2000); White (1789).
44. Davies (1992).
45. Wilkinson and Birkhead (1995).
46. Ekstrom et al. (2007).
47. Burkhardt et al. (2008: vol. 16 (1): 199).
48. Lesson (1831); Sushkin (1927); Bentz (1983).
49. Winterbottom et al. (2001).
50. Komisaruk et al. (2006, 2008).
51. Edvardsson and Arnqvist (2000).

4. TASTE

1. Darwin's (1871) idea of sexual selection comprised two parts: male–male competition and female choice. Darwin thought female choice largely responsible for the differences in plumage brightness between males and female. In contrast, male–male competition was responsible for differences in body size and weaponry. Hingston (1933), however, thought that bright colours might serve an intimidatory role and hence have evolved through male–male competition. Baker and Parker (1979) consider this idea illogical.

2. From the Darwin Correspondence – Burkhardt et al. (2008).

3. Weir (1869, 1870), see Burkhardt et al. (2008: 16 (2): 1175) and Burkhardt et al. (2009: 17: 115–16); C. Wiklund, personal communication (2009); Järvi et al. (1981); Wiklund and Järvi (1982). There is another intriguing example of birds having a sense of taste: the Greek writer Thucydides provides an account of an unusual strain of bubonic plague that hit Athens around 400 BC. Thucydides tells us that, in contrast to other plagues, carrion-eating birds avoided eating the unburied bodies, and when they did so they died. While hardly concrete evidence, it does suggest a sense of taste or smell and possibly a rapid learning ability (J. Mynott, personal communication).

4. Newton (1896); del Hoyo et al. (1992: vol. 1).

5. Malpighi (1665); Bellini (1665); Witt et al. (1994).

6. Rennie (1835). Montagu (1802) was an ornithologist; Johann Friedrich Blumenbach (1752–1840) was an anthropologist and anatomist famous for his anatomical study of the duck-billed platypus. Blumenbach (1805 – English translation, 1827, p. 260).

7. Newton (1896): his view was probably shaped by Friedrich Merkel, the great German anatomist, who in 1880 stated categorically that birds were devoid of taste buds. This was very odd for taste buds were already known to occur in fishes, amphibia, reptiles and mammals. Frustratingly, Newton gives no references so it isn't clear whether he had read Merkel, although it seems likely he knew of him.

8. Moore and Elliot (1946).

9. Berkhoudt (1980; 1985) and H. Berkhoudt, personal communication.

10. Botezat (1904); Bath (1906).

11. Berkhoudt (1985).

12. Brooker et al. (2008).

13. Rensch and Neunzig (1925).

14. Hainsworth and Wolf (1976); Mason and Clark (2000); van Heezik et al. (1983).

15. Jordt and Julius (2002); Birkhead (2003).

16. Kare and Mason (1986).
17. Beehler (1986); Majnep and Bulmer (1977).
18. J. Dumbacher, personal communication.
19. J. Dumbacher, personal communication.
20. Dumbacher et al. (1993). A video clip of Jack Dumbacher talking about his work is at http://www.calacademy.org/science/heroes/jdumbacher/
21. Audubon (1831–9).
22. Escalante and Daly (1994) cite an account of the flora and fauna of the Aztec world (pre-Columbian Mexico) (dated 1540–85) that mentions an inedible red bird that 'seems to correspond to the red warbler *Ergaticus ruber*'. Escalante and Daly (1994) extracted alkaloids from the birds' feathers.
23. Cott (1940); see also Anon (1987).
24. Cott (1947).
25. Cott (1945).
26. Cott picked out two people for particular praise: 'Col. R. Meinertzhagen DSO and Mr B. Vesey-Fitzgerald . . . both of whom have furthered the inquiry with many original observations of the greatest interest and relevance.' Oh dear! I wonder whether Cott may have been lead astray by these two. As later became clear, Meinertzhagen's entire life was a lie; one recent biography describing him as a colossal fraud. He was a pathological attention-seeker, and everything Meinterzhagen did, said or wrote was designed to bolster his self-image (Garfield, 2007). Brian Vesey-Fitzgerald was not entirely trustworthy either. Editor of the *Field*, and an extraordinarily prolific writer of natural history books, including the 1950s Ladybird children's books on birds, Vesey-Fitzgerald was exposed in 1949 as plagiarist by the eminent ornithologist the Reverend Peter H. T. Hartley (Hartley, 1947). That Vesey-Fitzgerald was not held in high esteem by other ornithologists was clear when one described him to me as 'a huntin' shootin' fishin' gasbag'.

5. SMELL

1. João dos Santos's account is cited in Friedmann (1955).
2. Audubon (1831–9); Audubon must here be referring to Richard Owen's dissection of the turkey vulture.
3. Gurney (1922: 240).
4. Audubon (1831–9).
5. In fact, Chapman did have a few reservations for he had noticed that with mallard ducks the direction of approach *did* matter – the difficulty, of course, was that, despite these hunters' statements to the contrary, they could not exclude vision and hearing. Elliot Coues (1842–99) was a surgeon and ornithologist.
6. This was one way that researchers attempted to verify their results in the late 1700s – not always with success (see Schickore, 2007: 43).
7. The debate was carried out in *Loudon's Magazine* (Gurney, 1922). While Waterton was in Guiana he taught one of his uncle's slaves, John Edmonstone, his skills. Edmonstone, by then freed and practising taxidermy in Edinburgh, in turn taught the teenage Charles Darwin to skin birds.
8. Confirmed by anatomy of olfactory cavity in the two species (see Bang 1960, 1965, 1971); Stager (1964, 1967).
9. Asafoetida, also known as devil's dung, is a powerful-smelling substance obtained from the umbelliferous plant *Ferula asafoetida*, and used – in tiny amounts – to flavour Worcestershire Sauce, and used by hunters as a lure! It is also used in enemas, and is a folk remedy for childhood diseases: Hill (1905).
10. Pickcheese – Gurney (1922); varied tit (Koyama, 1999; S. Koyama, personal communication); to 'fye out' is from Gurney (1922).
11. Tomalin (2008): Hardy used real events in his stories.
12. This story was related in the *Wiltshire Archaeological Magazine* for 1873, vol. xviii, p. 299 – cited in Gurney (1922).
13. Gurney (1922: 234).

14. Owen (1837).

15. Gurney (1922: 277) refers to several anatomical studies.

16. Gurney (1922).

17. Citing Gurney (1922) as evidence.

18. Textbook classics like Pierre-Paul Grasse's *Traité de Zoology: Oiseaux* (1950) and Jock Marshall's *Comparative Physiology of Birds* (1961) reiterated the same negative view. Even the much more recent and wonderful *Handbook of Birds of the World* says that, with the exception of a handful of species, most birds have a poor sense of smell (del Hoyo et al., 1992).

19. Taverner (1942).

20. Singular = concha, but these structures are paired, one on each side of the nose.

21. Van Buskirk and Nevitt (2007); Jones and Roper (1997).

22. From her daughter Molly, cited in Nevitt and Hagelin (2009).

23. Wenzel (2007).

24. Their study mentions '108' species but they counted rock dove, *Columba livia*, and feral pigeon, and also *Columba livia*, as different species – which they are not.

25. Strictly, the ratio between the longest diameter of the olfactory bulb and the longest diameter of the ipsilateral cerebral hemisphere.

26. Bang and Cobb (1968).

27. Clark et al. (1993); see also Balthazart and Schoffeniels (1979): the current consensus seems to be that a large olfactory bulb does indicate a good sense of smell, but a small one doesn't necessarily mean the opposite. There is still much to find out.

28. Bang and Cobb (1968).

29. Stager (1964); Bang and Cobb (1968). Today, ethyl mercaptan is added to domestic gas to make it detectable in case of leaks.

30. Bang and Cobb (1968) built on the previous studies of Bumm (1883) and Turner (1891).

31. S. Healy, personal communication.

32. Methods for dealing with allometry in comparative studies are given in Harvey and Pagel (1991).

33. Verner and Willson (1966); see also Harvey and Pagel (1991).

34. Methods for dealing with phylogeny in comparative studies are given in Harvey and Pagel (1991).

35. Healy and Guilford (1990).

36. Healy and Guilford (1990) used Bang and Cobb (1968) and Bang (1971) – 124 species in total.

37. Corfield et al. (2008b).

38. Corfield (2009).

39. Steiger et al. (2008). The nine species used in this study were the blue tit, black coucal, brown kiwi, canary, galah, red jungle fowl, kakapo, mallard and snow petrel. Steiger et al. also suggest that while birds with both relatively large olfactory bulb regions and olfactory gene repertoires may have an excellent sense of smell, the opposite might not be true.

40. Fisher (2002).

41. Newton (1896).

42. Owen (1879).

43. Jackson (1999: 326).

44. Benham (1906).

45. Wenzel (1965).

46. Wenzel (1968, 1971): by today's standards, recording from just two birds seems inadequate, but that was how physiologists operated then.

47. Wenzel (1971).

48. Wenzel (1971).

49. Aldrovandi (1599–1603); Buffon (1770–83).

50. Montagu (1813).

51. Gurney (1922).

52. Bang and Cobb (1968).

53. Bang and Wenzel (1985).

54. B. Wenzel, personal communication.

55. Loye Miller (1874–1970).

56. Perhaps it should be scumming? Chumming is the term used to attract sharks or other fish while fishing; it involves throwing chopped-up bait or fish bits into the sea.

57. Wisby and Halser (1954).

58. Jouventin and Weimerskirch (1990).

59. Grubb (1972).

60. Hutchinson and Wenzel (1980).

61. G. Nevitt, personal communication.

62. G. Nevitt, personal communication.

63. Bonadonna et al. (2006).

64. Collins (1884).

65. Nevitt et al. (2008).

66. Fleissner et al. (2003); Falkenberg et al. (2010).

67. Cited in Friedmann (1955).

6. A MAGNETIC SENSE

1. Gill et al. (2009).

2. Both Lockley and Lack must have been familiar with some early displacement studies of terns in the Caribbean conducted in the early 1900s by Watson (1908) and Watson and Lashley (1915); see also Wiltschko and Wiltschko (2003). The story of Caroline is told in Lockley (1942).

3. Lockley (1942).

4. Brooke (1990).

5. Brooke (1990).

6. Guilford et al. (2009).

7. Migratory restlessness is also known as *Zugunruhe*, because it was thought to be discovered by German ornithologists: it wasn't. It was discovered by an anonymous Frenchman: see Birkhead (2008).

8. Birkhead (2008); there have since been modifications of the basic design.

9. Middendorf (1859); Viguier (1882). The earth is a giant magnet with 'field lines' leaving the earth at its southern pole, re-entering again at the northern pole. At the equator the field lines are parallel to the earth's surface, but become steeper towards the poles. The strength (intensity) of the magnetic field also varies predictably across the earth's surface. Together, the angle of the field lines and the intensity of the magnetic field create unique 'magnetic signatures' for particular locations, which animals with magnetic maps can potentially utilise to ascertain their location. In the 1980s Robin Baker, then at Manchester University, conducted experiments on undergraduates that, to him at least, suggested a magnetic sense, although the rest of the scientific community were not convinced.

10. Thomson (1936).

11. Griffin (1944).

12. It is more complex than this – birds use both stars and the magnetic fields: Wiltschko and Wiltschko (1991).

13. Lohmann (2010).

14. Lohmann (2010).

15. Wilstchko and Wiltschko (2005); Fleissner et al. (2003); Falkenberg et al. (2010).

16. Ritz et al. (2000).

17. The dual receptor hypothesis is controversial, not all biologists accept it, and the mechanisms are hypothetical – so far.

7. EMOTIONS

1. Darwin (1871); Skutch (1996: 41); Gardiner (1832).

2. Predator distraction displays were recorded by Aristotle: see Armstrong (1956).

3. Tinbergen (1951); McFarland (1981: 151); see also Hinde (1966, 1982).

4. Griffin (1992) introduced the term 'cognitive ethology' – the start of a new field.

5. Gadagkar (2005).

6. Singer (1975); Dunbar and Shultz (2010).

7. The other two are: how did the universe begin? And, how did life begin? We have a reasonable idea about these two, but as far as consciousness is concerned, we have barely begun. For an up-to-date overview of consciousness in humans see Lane (2009).

8. Rolls (2005); Paul et al. (2005); Cabanac (1971).

9. Jessica Meade, my field assistant on Skomer in 2007, saw one of our colour-ringed guillemots killed by a great black-backed gull. This was a year with few rabbits – the gulls' normal prey.

10. Birkhead and Nettleship (1984).

11. K. Ashbrook, personal communication.

12. Ashbrook et al. (2008); M. P. Harris, personal communication.

13. Gould (1848).

14. Heinsohn (2009).

15. Tinbergen (1953).

16. Cockrem and Silverin (2002).

17. Cockrem (2007).

18. Schuett and Grober (2000).

19. Carere et al. (2001).

20. Bentham (1798).

21. Braithwaite (2010: 78).

22. J. Cockrem, personal communication.

23. Bolhuis and Giraldeau (2005).

24. *Sunday Times* (London), 14 December 2006.

25. Gentle and Wilson (2004). The younger chicks are when their beak is trimmed the more rapidly they recover and the less pain they seem to suffer, so trimming at one day old is common practice. Alternative methods of beak-trimming using infra-red heat cause less pain, and

in some areas there are moves to have beak trimming banned. See: http://www.poultryhub.org/index.php/Beak_trimming

26. Certain drugs do the same.

27. Young and Wang (2004).

28. E. Adkins-Regan, personal communication.

29. Zeki (2007): 'Judged by the world literature of love, romantic love has at its basis, a concept – that of unity, a state in which, at the height of passion, the desire of lovers is to be united to one another and to dissolve all distance between them. Sexual union is as close as humans can get to achieving that unity.'

30. Lack (1968); Birkhead and Møller (1992).

31. Dunbar and Shultz (2010); Dunbar (2010).

32. Harrison (1965).

33. Nelson (1978: 111).

34. Catchpole and Slater (2008).

35. Brown et al. (1988). Other co-operative breeding birds perform group displays, including communal dust bathing seen in white-winged choughs, and a remarkable dawn and dusk 'group dance' in Arabian babblers in which 'the birds press against each other, squeezing under, over and between partners in a curious frenzy of activity'.

36. Keverne et al. (1989); see also Dunbar (2010); I. Pepperberg, personal communication; she was keen to stress that these were anecdotal observations only.

37. Cabanac (1971).

POSTSCRIPT

1. Heppner (1965).

2. Montgomerie and Weatherhead (1997).

3. Simmons et al. (1988).

4. Gould, S. J. (1985).

5. Rosenblum (2010).

6. Chickens experiments: Pizzari et al. (2003); the increase in sperm numbers with a novel female is known as the Coolidge effect (after the American president Calvin Coolidge): the story is that 'the President and Mrs Coolidge were being shown [separately] around an experimental government farm. When [Mrs Coolidge] came to the chicken yard she noticed that a rooster was mating very frequently. She asked the attendant how often that happened and was told, "Dozens of times each day." Mrs Coolidge said, "Tell that to the President when he comes by." Upon being told, President asked, "Same hen every time?" The reply was, "Oh, no, Mr President, a different hen every time." President: "Tell that to Mrs Coolidge"' (Dewsbury 2000). Turnstone: Whitfield (1987); pigeons: Jitsumori et al. (1999).

7. Roseblum (2010).

8. Roseblum (2010).

Bibliography

Aldrovandi, U., 1599–1603, *Ornithologiae hoc est de avibus historiae*, Bologna, Italy.

Alvarez del Toro, M., 1971, 'On the biology of the American finfoot in southern Mexico', *Living Bird*, 10, 79–88.

Andersson, M. B., 1994, *Sexual Selection*, Princeton, NJ: Princeton University Press.

Anon., 1987. Obituary, H. B. Cott (1900–1987), Selwyn College Calendar 1987, 64–8.

Ansley, H., 1954, 'Do birds hear their songs as we do?', *Proceedings of the Linnaean Society of New York*, 63–5, 39–40.

Armstrong, E. A., 1956, 'Distraction display and the human predator', *The Ibis*, 98, 641–54.

Ashbrook, K., Wanless, S., Harris, M. P., and Hamer, K. C., 2008, 'Hitting the buffers: conspecific aggression undermines benefits of colonial breeding under adverse conditions', *Biology Letters*, 4, 630–33.

Audubon, J. J., 1831–9, *Ornithological Biography, or, an Account of the Habits of the Birds of the United States of America*, Edinburgh: A. Black.

Baker, R. R., and Parker, G. A., 1979, 'The evolution of bird colouration', *Philosophical Transactions of the Royal Society of London B*, 287, 67–130.

Baldner, L., 1666, *Vogel-, Fisch und Thierbuch*, unpublished MS, addl MSS 6485 and 6486, London, British Library.

— 1973, *Vogel-, Fisch und Thierbuch*, Einfurhrung von R. Lauterborn, Stuttgart: Muller und Schindler [facsimile edition].

Ballentine, B., and Hill, G. E., 2003, 'Female mate choice in relation to structural plumage coloration in blue grosbeaks', *The Condor*, 105, 593–8.

Bang, B. G., 1960, 'Anatomical evidence for olfactory function in some species of birds', *Nature*, 188, 547–9.

— 1965, 'Anatomical adaptations for olfaction in the snow petrel', *Nature*, 205, 513–15.

— 1971, 'Functional anatomy of the olfactory system in 23 orders of birds', *Acta Anatomica Supplementum*, 58, 1–76.

Bang, B. G., and Cobb, S., 1968, 'The size of the olfactory bulb in 108 species of birds', *The Auk*, 85, 55–61.

Bang, B. G., and Wenzel, B. M., 1985, 'Nasal cavity and olfactory system', in *Form and Function in Birds* (ed. King, A. S., and McLelland, J.), pp. 195–225, London: Academic Press.

Bath, W., 1906, 'Die Geschmacksorgane der Vögel; und Krokodile', *Arch. für Biontologie*, 1, 5–47.

Beehler, B. M., Pratt, T. K., and Zimmerman, D. A., 1986, *Birds of New Guinea*, Princeton, NJ: Princeton University Press.

Bellini, L., 1665, *Gustus Organum*, Bologna: Typis Pisarrianis.

Benham, W. B., 1906, 'The olfactory sense in *Apteryx*', *Nature*, 74, 222–3.

Bentham, J., 1798, *An Introduction to the Principles of Morals and Legislation*, London: T. Payne.

Bentz, G. D., 1983, 'Myology and histology of the phalloid organ of the buffalo weaver (*Bubalornis albirostris*)', *The Auk*, 100, 501–4.

Berkhoudt, H., 1980, 'Touch and taste in the mallard (*Anas platyrhynchos* L.)', PhD Thesis, University of Leiden.

— 1985, 'Structure and function of avian taste receptors', in *Form and Function in Birds*, vol. 3 (ed. King, A. S., and McLelland, J.), pp. 463–96, London: Academic Press.

Best, E., 2005, *Forest Lore of the Maori*, Wellington: Te Papa Press.

Birkhead, T. R., 2003, *The Red Canary*, London: Weidenfeld & Nicolson.

— 2008, *The Wisdom of Birds*, London: Bloomsbury.

Birkhead, T. R., and Møller, A. P., 1992, *Sperm Competition in Birds: Evolutionary Causes and Consequences*, London: Academic Press.

Birkhead, T. R., and Nettleship, D. N., 1984, 'Alloparental care in the common murre', *Canadian Journal of Zoology*, 62, 2121–4.

Blumenbach, J. F., 1827, *A Manual of Comparative Anatomy*, London: W. Simpkin & R. Marshall.

Bolhuis, J. J., and Giraldeau, L.-A., 2005, *The Behaviour of Animals: Mechanisms, Function and Evolution*, Hoboken, NJ: Wiley-Blackwell.

Bonadonna, F., Caro, S., Jouventin, P., and G. A. Nevitt, 2006, 'Evidence that blue petrel, *Halobaena caerulea*, fledglings can detect and orient to dimethyl sulphide', *Journal of Experimental Biology*, 209, 2165–9.

Botezat, E., 1904, 'Geschmacksorgane und andere nervöse Endapparate im Schnabel der Vögel (vorläufige Mitteilung)', *Biologisches Zentralblatt*, 24, 722–36.

Braithwaite, V. A., 2010, *Do Fish Feel Pain?*, Oxford: Oxford University Press.

Bray, C. W., and Thurlow, W. R., 1942, 'Interference and distortion in the cochlear responses of the pigeon', *Journal of Comparative Psychology*, 33, 279–89.

Brooke, M. de L., 1985, 'The effect of allopreening on tick burdens of molting Eudyptid penguins', *The Auk*, 102, 893–5.

— 1990, *The Manx Shearwater*, London: T. and A. D. Poyser.

Brooker, R. J., Widmaier, E. P., Graham, L. E., and Stiling, P. D., 2008, *Biology*, Boston, MA: McGraw-Hill.

Browne, E. D., 1988, 'Song sharing in a group-living songbird, the Australian magpie *Gymnorhina tibicen*. I. Vocal sharing within and among groups', *Behaviour*, 104, 1–28.

Brumm, H., 2004, 'The impact of environmental noise on song amplitude in a territorial bird', *Journal of Animal Ecology*, 73, 434–40.

— 2009, 'Song amplitude and body size in birds', *Behavioural Ecology & Sociobiology*, 63, 1157–65.

Buffon, G.-L., 1770–83, *Histoire Naturelle des Oiseaux*, Paris.

Buller, S. W. L., 1873, *A History of the Birds of New Zealand*, London: J. Van Voorst.

Bumm, A., 1883, 'Das Großhirn der Vögel', *Z wiss Zool*, 38, 430–67.

Burkhardt, F., Secord, J. A., Dean, S. A., Evans, S., Innes, S., Pearn, A. M., and White, P., 2008, *The Correspondence of Charles Darwin*, vol. 16, *1868, Parts 1 and 2*, Cambridge: Cambridge University Press.

— 2009, *The Correspondence of Charles Darwin*, vol. 17, *1869*, Cambridge: Cambridge University Press.

Burton, R. F., 2008, 'The scaling of eye size in adult birds: relationship to brain, head and body sizes', *Vision Research*, 48, 2345–51.

Cabanac, M., 1971, 'Physiological role of pleasure', *Science*, 173, 1103–7.

Carere, C., Welink, D., Drent, P. J., Koolhaas, J. M., and Groothius, T. G., 2001, 'Effect of social defeat in a territorial bird (*Parus major*) selected for different coping styles', *Physiological Behavior*, 73, 427–33.

Carvell, G. E., and Simmons, D. J., 1990, 'Biometric analyses of vibrissal tactile discrimination in the rat', *Neuroscience*, 10, 2638–48.

Catchpole, C. K., and Slater, P. J. B., 2008, *Bird Song: Themes and Varations*, 2nd edn, Cambridge: Cambridge University Press.

Clark, L., Avilova, K. V., and Bean, N. J., 1993, 'Odor thresholds in passerines', *Comparative Biochemistry and Physiology A*, 104, 305–12.

Cobb, N., 1915, 'Nematodes and their relationships', *Yearbook of the United States Department of Agriculture 1914*, pp. 457–90. Washington, DC: Dept of Agriculture.

Cockrem, J. F., 2007, 'Stress, corticosterone responses and avian personalities', *Journal of Ornithology*, 148 (Suppl. 2), S169–S178.

Cockrem, J. F., and Silverin, B., 2002, 'Sight of a predator can stimulate a corticosterone response in the great tit (*Parus major*)', *General and Comparative Endocrinology*, 125, 248–55.

Coiter, V., 1572, *Externarum et internarum principalium humani corporis partium tabulae*, Nuremberg: In officina Theodorici Gerlatzeni.

Cole, F. J., 1944, *A History of Comparative Anatomy from Aristotle to the Eighteenth Century*, London: Macmillan.

Collins, J. W., 1884, 'Notes on the habits and methods of capture of various species of seabirds that occur on the fishing banks off the eastern coast of North America', *Report of the Commissioner of Fish and Fisheries for 1882*, 13, 311–35.

Collins, S., 2000, 'Men's voices and women's voices', *Animal Behaviour*, 60, 773–80.

Corfield, J. R., 2009, 'Evolution of the brain and sensory systems of the kiwi', unpublished PhD Thesis, University of Auckland.

Corfield, J. R., Wild, J. M., Hauber, M. E., Parsons, S., and Kubke, M. F., 2008a, 'Evolution of brain size in the palaeognath lineage, with an emphasis on New Zealand ratites', *Brain, Behaviour & Evolution*, 71, 87–99.

Corfield, J. R., Wild, J. M., Cowan, B. R., Parsons, S., and Kubke, M. F., 2008b, 'MRI of postmortem specimens of endangered species for comparative brain anatomy', *Nature Protocols*, 3, 597–605.

Cott, H. B., 1940, *Adaptive Colouration in Animals*, London: Methuen & Co.

— 1945, 'The edibility of birds', *Nature*, 156, 736–7.

— 1947, 'The edibility of birds: illustrated by five years' experiments and observations (1941–1946) on the food preferences of the hornet, cat and man; and considered with special reference to the theories of adaptive coloration', *Proceedings of the Zoological Society of London*, 116, 371–524.

Cunningham, S. J., Castro, I., and M. Alley, 2007, 'A new prey-detection mechanism for kiwi (*Apteryx* spp.) suggests convergent evolution between paleognathous and neognathous birds', *Journal of Anatomy*, 211, 493–502.

Cuthill, I. C., 2006, 'Colour perception', in *Bird Coloration: Mechanisms and Measurements* (ed. Hill, G. E., and McGraw, K.), pp. 3–40. Cambridge, MA: Harvard University Press.

Darwin, C., 1871, *The Descent of Man, and Selection in Relation to Sex*, London: J. Murray.

Davies, N. B., 1992, *Dunnock Behaviour and Social Evolution*, Oxford: Oxford University Press.

— 2000, *Cuckoos, Cowbirds and other Cheats*, London: Poyser.

Dawkins, M. S., 2006, 'Through animal eyes: what behaviour tells us', *Applied Animal Behaviour Science*, 100, 4–10.

Defoe, D., 1724–6, *A Tour Through The Whole Island of Great Britain*, London.

Derham, W., 1713, *Physico-theology*. London: W. and J. Innys.

Dewsbury, D. A., 2000, 'Frank A. Beach, master teacher', *Portraits of Pioneers in Psychology*, 4, 269–81.

Dijkgraaf, S., 1960, 'Spallanzani's unpublished experiments on the sensory basis of object perception in bats', *Isis*, 51, 9–20.

Dooling, R. J., Fay, R. R., and Popper, A. N., 2000, *Comparative Hearing: Birds and Reptiles*, New York: Springer-Verlag.

Draganoiu, T. I., Nagle, L., and Kreutzer, M., 2002, 'Directional female preference for an exaggerated male trait in canary (*Serinus canaria*) song', *Proc. R. Soc. Lond. B*, 269, 2525–31.

Drent, R., 1975, 'Incubation', in *Avian Biology* (ed. Farner, D. S., and King, J. R.), pp. 333–420, New York: Academic Press.

Dumbacher, J. P., Beehler, B. M., Spande, T. F., Garraffo, H. M., and Daly, J. W., 1993, 'Pitohui: how toxic and to whom?', *Science*, 259, 582–3.

Dunbar, R. I. M., 2010, 'The social role of touch in humans and primates: Behavioural function and neurobiological mechanisms', *Neuroscience and Biobehavioural Reviews*, 34, 260–68.

Dunbar, R. I. M., and Shultz, S., 2010, 'Bondedness and sociality', *Behaviour*, 147, 775–803.

Eaton, M. D., and Lanyon, S. M., 2003, 'The ubiquity of avian ultraviolet plumage reflectance', *Proc. R. Soc. Lond. B*, 270, 1721–6.

Edvardsson, M., and Arnqvist, G., 2000, 'Copulatory courtship and cryptic female choice in red flour beetles *Tribolium castaneum*', *Proc. R. Soc. Lond. B*, 267, 446–8.

Ekstrom, J. M. M., Burke, T., Randrianaina, L., and Birkhead, T. R., 2007, 'Unusual sex roles in a highly promiscuous parrot: the greater vasa parrot *Caracopsis vasa*', *The Ibis*, 149, 313–20.

Escalante, P., and Daly, J. W., 1994, 'Alkaloids in extracts of feathers of the red warbler', *Journal of Ornithology*, 135, 410.

Falkenberg, G., Fleissner, G., Schuchardt, K., Kuehbacher, M., Thalau, P., Mouritsen, H., Heyers, D., Wellenreuther, G., and Fleissner, G., 2010, 'Avian magnetoreception: elaborate iron mineral-containing dendrites in the upper beak seem to be a common feature of birds', *PLoS ONE*, 5, 1–9.

Fenton, M. B., and Bell, G. P., 1981, 'Recognition of species of insectivorous bats by their echolocation calls', *Journal of Mammalogy*, 62: 233–43.

Fisher, C., 2002, *A Passion for Natural History: The Life and Legacy of the 13th Earl of Derby*, Liverpool: National Museums and Galleries, Merseyside.

Fitzpatrick, J. W. et al. (sixteen co-authors), 2005, 'Ivory-billed woodpecker (*Campephilus principalis*) persists in continental North America', *Science*, 308, 1460–62.

Fleissner, G., Holtkamp-Rotzler, E., Hanzlik, M., Winklhofer, M., Fleissner, G., Petersen, N., and Wiltschko, W., 2003, 'Ultrastructural analysis of a putative magnetoreceptor in the beak of homing pigeons', *The Journal of Comparative Neurology*, 458, 350–60.

Forstmeier, W., and Birkhead, T. R., 2004, 'Repeatability of mate choice in the zebra finch: consistency within and between females', *Animal Behaviour*, 68, 1017–28.

Fox, R., Lehmkuhle, S., and Westerndorf, D. H., 1976, 'Falcon visual acuity', *Science*, 192, 263–5.

Friedmann, H., 1955, 'The honey-guides', *Bulletin of the United States National Museum*, 208, 292.

Gadagkar, R., 2005, 'Donald Griffin strove to give animals their due', *Resonance*, 10, 3–5.

Galambos, R., 1942, 'The avoidance of obstacles by flying bats: Spallanzani's ideas (1794) and later theories,' *Isis*, 34, 132–40.

Gardiner, W., 1832, *The Music of Nature; or, An Attempt to Prove that what is Passionate and Pleasing in the Art of Singing, Speaking and Performing upon Musical Instruments, is Derived from Sounds of the Animated World*, London: Longman.

Garfield, B., 2007, *The Meinertzhagen Mystery*, Washington, DC: Potomac Books.

Gentle, M., and Wilson, S., 2004, 'Pain and the laying hen', in *Welfare of the Laying Hen* (ed. Perry, G. C.), pp. 165–75, Wallingford: CABI.

Gerritsen, A. F. C., Van Heezik, Y. M., and Swennen, C., 1983, 'Chemoreception in two further *Calidris* species (*C. maritima and C. canutus*) with comparison of the relative importance of chemoreception during foraging in *Calidris* species', *Netherlands Journal of Zoology*, 33, 485–96.

Gill, R. E., Tibbits, T. L., Douglas, D. C., Handal, C. M., Mulcahy, D. M., Gottschlack, J. C., Warnock, N., McCafferey, B. J., Battley, P. F., and Piersma, T., 2009, 'Extreme endurance flights by landbirds crossing the Pacific Ocean: ecological corridor rather than barrier?', *Proc. R. Soc. Lond. B*, 276, 447–57.

Gilliard, E. T., 1962, 'On the breeding behaviour of the Cock-of-the-Rock (Aves, *Rupicola rupicola*)', *Bulletin of the American Museum of Natural History*, 124, 31–68.

Goujon, D. E., 1869, 'An apparatus of tactile corpuscles situated in the beaks of parrots', *Journal de l' Anatomie et de la Physiologie Normales et Pathologiques de l'Homme*, 6, 449–55.

Gould, J., 1848, *The Birds of Australia*, London: published by the author, seven volumes.

Gould, S. J., 1985, *The Flamingo's Smile: Essays in Natural History*, New York: W. W. Norton & Co.

Grassé, P. P., 1950, *Traité de Zoologie: Oiseaux*, Paris: Masson.

Griffin, D. R., 1944, 'The sensory basis of bird navigation', *The Quarterly Review of Biology*, 19, 15–31.

— 1958, *Listening in the Dark: The Acoustic Orientation of Bats and Men*, New Haven, Conn.: Yale University Press.

— 1976, *The Question of Animal Awareness: Evolutionary Continuity of Mental Experience*, New York, NY: The Rockefeller University Press.

— 1992, *Animal Minds*, Chicago IL: University of Chicago Press.

Grubb, T., 1972, 'Smelling and foraging in petrels and shearwaters', *Nature*, 237, 404–5.

Guilford, T., Meade, J., Willis, J., Phillips, R. A., Boyle, D., Roberts, S., Collett, M., Freeman, R., and Perrins, C. M., 2009, 'Migration and stop-over in a small pelagic seabird, the Manx shearwater *Puffinus puffinus*: insights from machine learning', *Proc. R. Soc. Lond. B*, 276, 1215–23.

Gurney, J. H., 1922, 'On the sense of smell possessed by birds', *The Ibis*, 2, 225–53.

Hainsworth, F. R., and Wolf, L. L., 1976, 'Nectar characteristics and food selection by hummingbirds', *Oecologica*, 25, 101–13.

Handford, P., and Nottebohm, F., 1976, 'Allozymic and morphological variation in population samples of rufous-collared sparrow, *Zonotrichia capensis*, in relation to vocal dialects', *Evolution*, 30, 802–17.

Harris, L. J., 1969, Footedness in parrots: three centuries of research, theory, and mere surmise, *Canadian Journal of Psychology*, 43, 369–96.

Harrison, C. J. O., 1965, 'Allopreening as agonistic behaviour', *Behaviour*, 24, 161–209.

Harting, J. E., 1883, *Essays on Sport and Natural History*, London: Horace Cox.

Hartley, P. H. T., 1947, 'Review of *Background to Birds* by B. Vesey-Fitzgerald', *The Ibis*, 91, 539–40.

Harvey, P. H., and Pagel, M. D., 1991, *The Comparative Method in Evolutionary Biology*, Oxford: Oxford University Press.

Healy, S., and Guilford, T., 1990, 'Olfactory-bulb size and nocturnality in birds', *Evolution*, 44, 339–46.

Heinsohn, R., 2009, 'White-winged choughs: the social consequences of boom and bust', in *Boom and Bust: Bird Stories for a Dry Country* (ed. Robin, L., Heinsohn, R., and Joseph, L.), pp. 223–40, Victoria, Australia: CSIRO Publishing.

Henry, R., 1903, *The Habits of the Flightless Birds of New Zealand, with Notes on other New Zealand birds*, Wellington, NZ: Government Printer.

Heppner, F., 1965, 'Sensory mechanisms and environmental cues used by the American robin in locating earthworms', *Condor*, 67, 247–56.

Hill, A., 1905, 'Can birds smell?', *Nature*, 1840, 318–19.

Hill, G. E., 2007, *Ivorybill Hunters*, Oxford: Oxford University Press.

Hill, G. E., and McGraw, K. J., 2006, *Bird Coloration: Function and Evolution*, Cambridge, MA: Harvard University Press.

Hinde, R. A., 1966, *Animal Behaviour: A Synthesis of Ethology and Comparative Psychology*, Maidenhead: McGraw-Hill.

—— 1982, *Ethology*, Oxford: Oxford University Press.

Hingston, R. W. G., 1933, *The Meaning of Animal Colour and Adornment*, London: Edward Arnold.

Homberger, D. G., and de Silva, K. N., 2000, 'Functional microanatomy of the feather-bearing integument: implications for the evolution of birds and avian flight', *American Zoologist*, 40, 553–74.

Howland, H. C., Merola, S., and Basarab, J. R., 2004, 'The allometry and scaling of the size of vertebrate eyes', *Vision Research*, 44, 2043–65.

Hulse, S. H., MacDougall-Shackleton, S. A., and Wisniewski, A. B., 1997, 'Auditory scene analysis by songbirds: stream segregation of birdsong by European starlings (*Sturnus vulgaris*)', *Journal of Comparative Psychology*, 111, 3–13.

Hultcrantz, M., and Simonoska, R., 2006, 'Estrogen and hearing: a summary of recent investigations', *Acta Oto-Laryngologica*, 126, 10–14.

Hunter, M. L., and Krebs, J. R., 1979, 'Geographical variation in the song of the great tit (*Parus major*) in relation to ecological factors', *Journal of Animal Ecology*, 48, 759–85.

Hutchinson, L. V., and Wenzel, M., 1980, 'Olfactory guidance in foraging by Procellariiforms', *The Condor,* 82, 314–19.

Ings, S., 2007, *The Eye: A Natural History*, London: Bloomsbury.

Jackson, C. E., 1999, *Dictionary of Bird Artists of the World*, Woodbridge: Antique Collectors' Club.

Järvi, T., Sillén-Tullberg, B., and Wiklund, C., 1981, 'The cost of being aposematic: an experimental study of predation on larvae of *Papilio achaon* by the great tit *Parus major*', *Oikos*, 36, 267–72.

Jenner, E., 1788, 'Observations on the natural history of the cuckoo', *Philosophical Transactions of the Royal Society*, 78, 219–37.

Jitsumori, M., Natori, M., and Okuyama, K., 1999, 'Recognition of moving video images of conspecifics by pigeons: effects of individual static and dynamic motion cues', *Animal Learning & Behavior*, 27, 303–15.

Jones, D. N., and Goth, A., 2008, *Mound-builders*, Victoria, Australia: CSIRO Publishing.

Jones, R. B., and Roper, T. J., 1997, 'Olfaction in the domestic fowl: a critical review', *Physiology & Behavior*, 62, 1009–18.

Jordt, S. E., and Julius, D. 2002, 'Molecular basis for species-specific sensitivity to "hot" chili peppers', *Cell*, 108, 421–30.

Jouventin, P., and Weimerskirch, H., 1990, 'Satellite tracking of wandering albatrosses', *Nature*, 343, 746–8.

Kare, M. R., and Mason, J. R., 1986, 'Chemical senses in birds', in *Avian Physiology* (ed. Sturkie, P. D.), New York, NY: Springer Verlag.

Keverne, E. B., Martensz, N. D., and Tuite, B., 1989, 'Beta-endorphin concentrations in cerebrospinal fluid of monkeys are influenced by grooming relationships', *Psychoneuroendocrinology*, 14, 155–61.

Knox, A. G., 1983, 'Handedness in crossbills *Loxia* and the akepa *Loxops coccinea*', *Bulletin of the British Ornithologists' Club*, 103, 114–18.

Komisaruk, B. R., Beyer, C., and Whipple, B., 2008, 'Orgasm', *The Psychologist*, 21, 100–103.

Komisaruk, B. R., Beyer-Flores, C., and Whipple, B., 2006, *The Science of Orgasm*, Baltimore, MD: Johns Hopkins University Press.

Konishi, M., 1973, 'How the owl tracks its prey', *American Scientist*, 61, 414–24.

Konishi, M., and Knudsen, E. L., 1979, 'The oilbird: hearing and echolocation', *Science*, 204, 425–7.

Koyama, S., 1999, *Tricks Using Varied Tits: Its History and Structure* [in Japanese], Tokyo: Hosei University Press.

Krebs, J. R., and Davies, N. B., 1997, *Behavioural Ecology: An Evolutionary Approach*, 4th edn, Oxford: Blackwell.

Krulis, V., 1978, 'Struktur und Verteilung von Tastrezeptoren im Schnabel-Zungenbereich von Singvögeln im besonderen der Fringillidae', *Revue Suisse de Zoologie*, 85, 385–447.

Lack, D., 1956, *Swifts in a Tower*, London: Methuen.

— 1968, *Ecological Adaptations for Breeding in Birds*, London: Methuen.

Lane, N., 2009, *Life Ascending*, London: Profile Books.

Lea, R. B., and Klandorf, H., 2002, 'The brood patch', in *Avian Eggs and Incubation* (ed. Deeming, C.), pp. 156–89, Oxford: Oxford University Press.

Lesson, R. P., 1831, *Traite d'ornithologie*, Paris: Bertrand.

Lockley, R. M., 1942, *Shearwaters*, London: Dent.

Lohmann, K. J., 2010, 'Animal behaviour: magnetic-field perception', *Nature*, 464, 1140–42.

Lucas, J. R., Freeman, T. M., Long, G. R., and Krishnan, A., 2007, 'Seasonal variation in avian auditory evoked responses to tones: a comparative analysis of Carolina chickadees, tufted titmice, and white-breasted nuthatches', *Journal of Comparative Physiology A*, 193, 201–15.

Macdonald, H., 2006, *Falcon*, London: Reaktion Books.

Majnep, I. S., and Bulmer, R., 1977, *Birds of my Kalam Country*, Auckland, NZ: Auckland University Press.

Malpighi, M., 1665, *Epistolae Anatomicae de Cerebro ac Lingua*, Bologna, Italy: Typis Antonij Pisarrij.

Marler, P., 1959, 'Developments in the study of animal communication', in *Darwin's Biological Work* (ed. Bell, P. R.), pp. 150–206, Cambridge: Cambridge University Press.

Marler, P., and Slabbekoorn, H. W., 2004, *Nature's Music: The Science of Birdsong*, London: Academic Press.

Marshall, A. J., 1961, *Biology and Comparative Physiology of Birds*, New York, NY: Academic Press.

Martin, G., 1990, *Birds by Night*, London: Poyser.

Martin, G. R., and Osorio, D., 2008, 'Vision in birds', in *The Senses: A Comprehensive Reference* (ed. Basbaum, A. I., Kaneko, A., Shepherd, G. M., Westheimer, G., Albright, T. D., Masland, R. H., Dallos, P., Oertel, D., Firestein, D., Beauchamp, G. K., Bushnell, M. C., Kaas, J. C., and Gardner, E.), Berlin: Elsevier.

Martin, G. R., Wilson, K.-J., Wild, J. M., Parsons, S., Kubke, M. F., and Corfield, J., 2007, 'Kiwi forego vision in the guidance of their nocturnal activites', *PLoS ONE* 2 (2) e198, 1–6.

Mason, J. R., and Clark, L., 2000, 'The chemical senses of birds', in *Sturkie's Avian Physiology* (ed. Sturkie, P. D.), pp. 39–56, San Diego: Academic Press.

McCurrich, J. P., 1930, *Leonardo da Vinci: The Anatomist*, Washington, DC: Carnegie Institute, Washington.

McFarland, D., 1981, *The Oxford Companion to Animal Behaviour*, New York, NY: Oxford University Press.

Merton, D. V., Morris, R. B., and Atkinson, I. A. E., 1984, 'Lek behaviour in a parrot: the kakapo *Strigops habroptilus* of New Zealand', *The Ibis*, 126, 277–83.

Middendorf, A. V., 1859, 'Die Isepiptesen Rußlands', *Mémoires de l'Académie Impériale des Sciences de St. Pétersbourg*, VI, 1–143.

Mikkola, H., 1983, *Owls of Europe*, New York: T. & A. D. Poyser.

Miller, L., 1942, 'Some tagging experiments with back-footed albatrosses', *The Condor*, 44, 3–9.

Mockford, E. J., and Marshall, R. C., 2009, 'Effects of urban noise on song and response behaviour in great tits', *Proc. R. Soc. Lond. B*, 276, 2976–85.

Montagu, G., 1802, *Ornithological Dictionary*, London: White.

— 1813, *Supplement to the Ornithological Dictionary*, Exeter: Woolmer.

Montgomerie, R., and Birkhead, T. R., 2009, 'Samuel Pepys's hand-coloured copy of John Ray's "The Ornithology of Francis Willughby" (1678)', *J. Ornithol.*, 150, 883–91.

Montgomerie, R., and Weatherhead, P. J., 1997, 'How robins find worms', *Animal Behaviour*, 54, 143–51.

More, H., 1653, *An Antidote Against Atheism: Or an Appeal to the Natural Faculties of the Minds of Man, Whether there be not a God*, London: Daniel.

Morton, E. S., 1975, 'Ecological sources of selection on avian sounds', *American Naturalist*, 109, 17–34.

Nagel, T., 1974, 'What is it like to be a bat?', *The Philosophical Review*, 83, 435–50.

Naguib, M., 1995, 'Auditory distance assessment of singing conspecifics in Carolina wrens: the role of reverberation and frequency-dependent attenuation', *Animal Behaviour*, 50, 1297–307.

Necker, R., 1985, 'Observations on the function of a slowly-adapting mechanoreceptor associated with filoplumes in the feathered skin of pigeons', *Journal of Comparative Physiology A*, 156, 391–4.

Nelson, J. B., 1978, *The Gannet*, Berkhamsted: Poyser.

Nevitt, G. A., 2008, 'Sensory ecology on the high seas: the odor world of the procellariforme seabirds', *Journal of Experimental Biology*, 211, 1706–13.

Nevitt, G. A., and Hagelin, J. C., 2009, 'Olfaction in birds: a dedication to the pioneering spirit of Bernice Wenzel and Betsy Bang', *Annals of the New York Academy of Sciences*, 1170, 424–7.

Nevitt, G. A., Losekoot, M., and Weimerskirch, H., 2008, 'Evidence for olfactory search in wandering albatross, *Diomedea exulans*', *Proceedings of the National Academy of Sciences*, USA, 105, 4576–81.

Newton, A., 1896, *A Dictionary of Birds*, London: A. & C. Black.

Nilsson, D. E., and Pelger, S., 1994, 'A pessimistic estimate of the time required for an eye to evolve', *Proc. R. Soc. Lond. B*, 256, 53–8.

Nottebohm, F., 1977, 'Asymmetries in neural control of vocalization in the canary. In *Lateralisation in the Nervous System* (ed. Harnad, S., Doty, R. W., Goldstein, L., Jaynes, J., and Krauthamer, G.), New York, NY: Academic Press.

Novick, A., 1959, 'Acoustic orientation in the cave swiftlet', *Biological Bulletin*, 117, 497–503.

Owen, R., 1837, No title. *Proceedings of the Zoological Society of London*, 1837, 33–5.

— 1879, *Memoirs on the Extinct Wingless Birds of New Zealand. With an Appendix on Those in England, Australia, Newfoundland, Mauritius and Rodriguez*, London: John van Voorst.

Paley, W., 1802, *Natural Theology: Or Evidences of the Existence and Attributes of the Deity Collected from the Appearances of Nature*, London.

Parker, T. J., 1891, 'Observations on the anatomy and development of *Apteryx*', *Phil. Trans. R. Soc. London B*, 182, 25–134.

Paul, E. S., Harding, E. J., and Mendl, M., 2005, 'Measuring emotional processes in animals: the utility of a cognitive approach', *Neuroscience and Biobehavioural Reviews*, 29, 469–91.

Payne, R. S., 1971, 'Acoustic location of prey by barn owls', *Journal of Experimental Biology*, 54, 535–73.

Perrault, C., 1680, *Essais de physique, ou recueil de plusieurs traitez touchant les choses naturelles*, Paris: J. B. Coignard.

Pfeffer, K. von, 1952, 'Untersuchungen zur Morphologie und Entwicklung der Fadenfedern', *Zoologische Jahrbücher Abteilung für Anatomie*, 72, 67–100.

Piersma, T., van Aelst, R., Kurk, K., Berkhoudt, H., and Mass, L. R. M., 1998, 'A new pressure sensory mechanism for prey detection in birds: the use of principles of seabed dynamics?', *Proc. R. Soc. Lond. B*, 265, 1377–83.

Pizzari, T., Cornwallis, C. K., Lovlie, H., Jakobsson, S., and Birkhead, T. R., 2003, 'Sophisticated sperm allocation in male fowl', *Nature*, 426, 70–74.

Pumphrey, R. J., 1948, 'The sense organs of birds', *The Ibis*, 90, 171–99.

Radford, A. N., 2008, 'Duration and outcome of intergroup conflict influences intragroup affiliative behaviour', *Proc. R. Soc. Lond. B*, 275, 2787–91.

Rattenborg, N. C., Amlaner, C. J., and Lima, S. L., 2000, 'Behavioral, neurophysiological and evolutionary perspectives on unihemispheric sleep', *Neuroscience and Biobehavioural Reviews*, 24, 817–42.

Rattenborg, N. C., Lima, S. L., and Amlaner, C. J., 1999, 'Facultative control of avian unihemispheric sleep under the risk of predation', *Behavioural Brain Research*, 105, 163–72.

Ray, J., 1678, *The Ornithology of Francis Willughby*, London: John Martyn.

Rennie, J., 1835, *The Faculties of Birds*, London: Charles Knight.

Rensch, B., and Neunzig, R., 1925, 'Experimentelle Untersuchungen uber den Geschmackssinn der Vögel II', *Journal of Ornithology*, 73, 633–46.

Ritz, T., Adem, S., and Schulten, K., 2000, 'A model for vision-based magnetoreception in birds', *Biophysical Journal*, 78, 707–18.

Rochon-Duvigneaud, A., 1943, *Les yeux et la vision des Vertébrés*, Paris: Masson.

Rogers, L. J., 1982, 'Light experience and asymmetry of brain function in chickens', *Nature*, 297, 223–5.

— 2008, 'Development and function of lateralization in the avian brain', *Brain Research Bulletin*, 76, 235–44.

Rogers, L. J., Zucca, P., and Vallortigara, G., 2004, 'Advantages of having a lateralized brain', *Proc. R. Soc. Lond. B*, 271, S420–S422.

Rolls, E. T., 2005, *Emotions Explained*, Oxford: Oxford University Press.

Rosenblum, L. D., 2010, *See What I'm Saying*, New York, NY: Horton.

Ruestow, E. G., 1996, *The Microscope in the Dutch Republic*, Cambridge: Cambridge University Press.

Sade, J., Handrich, Y., Bernheim, J., and Cohen, D., 2008, 'Pressure equilibration in the penguin middle ear', *Acta Oto-Laryngologica*, 128, 18–21.

Sayle, C. E., 1927, *The Works of Sir Thomas Browne*, Edinburgh: Grant.

Schickore, J., 2007, *The Microscope and the Eye: a History of Reflections, 1740–1870*, Chicago, IL: University of Chicago Press.

Schlegel, H., and Wulvergorst, A. H. V., 1844–53, *Traite de Fauconnerie*, Leiden: Arnz.

Schuett, G. W., and Grober, M. S., 2000, 'Post-fight levels of plasma lactate and corticosterone in male copperheads *Agkistrodon contortrix* (Serpentes, Viperidae): differences between winners and losers', *Physiology & Behavior*, 71, 335–41.

Schwartzkopff, J., 1949, 'Über den Sitz und Leistung von Gehör und Vibrationssinn bei Vögeln', *Zeitschrift für vergleichende Physiologie*, 31, 527–608.

Senevirante, S. S., and Jones, I. L., 2008, 'Mechanosensory function for facial ornamentation in the whiskered auklet, a crevice-dwelling seabird', *Behavioural Ecology*, 19, 184–790.

— 2010, 'Origin and maintenance of mechanosensory feather ornaments', *Animal Behaviour*, 79, 637–44.

Sibley, D., 2000, *The Sibley Guide to Birds*, New York, NY: Alfred A. Knopf.

Simmons, R. L., Barnard, P., and Jamieson, I. G., 1998, 'What precipitates influxes of wetland birds to ephemeral pans in arid landscapes? Observations from Namibia', *Ostrich* 70, 145–8.

Singer, P., 1975, *Animal Liberation*, New York, NY: Avon Books.

Skutch, A. F., 1935, 'Helpers at the nest', *The Auk*, 52, 257–73.

— 1996, *The Minds of Birds*, College Station, TX: Texas A&M Press.

Slabbekoorn, H., and Peet, M., 2003, 'Birds sing at a higher pitch in urban noise', *Nature* 424, 267.

Slonaker, J. R., 1897, *A Comparative Study of the Area of Acute Vision in Vertebrates*, Cambridge, MA.: Harvard University Press.

Snyder, A. W., and Miller, W. H., 1978, 'Telephoto lens system of falconiform eyes', *Nature*, 275, 127–9.

Stager, K. E., 1964, 'The role of olfaction in food location by the turkey vulture (*Cathartes aura*)', *Los Angeles County Museum Contributions in Science*, 81, 547–9.

— 1967, 'Avian olfaction', *American Zoologist*, 7, 415–20.

Steiger, S. S., Fidler, A. E., Valcu, M., and Kempenaers, B., 2008, 'Avian olfactory receptor gene repertoires: evidence for a well-developed sense of smell in birds?' *Proc. R. Soc. Lond. B*, 275, 2309–17.

Stetson, C., Fiesta, M. P., and Eagleman, D. M., 2007, 'Does time really slow down during a frightening event?', *PLoS ONE*, 2, e1295.

Stowe, M., Bugnyar, T., Schloegl, C., Heinrich, B., Kotrschal, K., and Mostl, E., 2008, 'Corticosterone excretion patterns and affiliative behavior over development in ravens (*Corvus corax*)', *Hormones and Behaviour*, 53, 208–16.

Sushkin, P. P., 1927, 'On the anatomy and classification of the weaver birds', *Bulletin of the American Museum of Natural History*, 57, 1–32.

Swaddle, J. P., Ruff, D. A., Page, L. C., Frame, A. M., and Long, V. C., 2008, 'Test of receiver perceptual performance: European starlings' ability to detect asymmetry in a naturalistic trait', *Animal Behaviour*, 76, 487–95.

Taverner, P. A., 1942, 'The sense of smell in birds', *The Auk*, 59, 462–3.

Thomson, A. L., 1936, *Bird Migration: A Short Account*, London: H. F. & G. Witherby.

— 1964, *A New Dictionary of Birds*, London and Edinburgh: Thomas Nelson & Sons.

Thorpe, W. H., 1961. *Bird-Song*, Cambridge: Cambridge University Press.

Tinbergen, N., 1951, *The Study of Instinct*, Oxford: Clarendon Press.

— 1963, 'On aims and methods of ethology', *Zeitschrift für Tierpsychologie*, 20, 410–33.

Tomalin, C., 2008, *Thomas Hardy. The Time-Torn Man*, London: Penguin Books.

Tryon, C. A., 1943, 'The great grey owl as a predator on pocket gophers', *The Wilson Bulletin*, 55, 130–31.

Tucker, V. A., 2000, 'The deep fovea, sideways vision and spiral flight paths in raptors', *Journal of Experimental Biology*, 203, 3745–54.

Tucker, V. A., Tucker, A. E., Akers, K., and Enderson, J. H., 2000, 'Curved flight paths and sideways vision in peregrine falcons (*Falco peregrinus*)', *Journal of Experimental Biology*, 203, 3755–63.

Turner, C. H., 1891, 'Morphology of the avian brains. I – Taxonomic value of the avian brain and the histology of the cerebrum', *The Journal of Comparative Neurology*, 1, 39–92.

Vallet, E. M., Kreutzer, M. L., Beme, I., and Kiosseva, L., 1997, '"Sexy" syllables in male canary songs: honest signals of motor constraints on male vocal production?' *Advances in Ethology*, 32, 132.

Van Buskirk, R. W., and Nevitt, G. A., 2007, 'Evolutionary arguments for olfactory behavior in modern birds', *ChemoSense*, 10, 1–6.

Van Heezik, Y. M., Gerritsen, A. F. C., and Swennen, C., 1983, 'The influence of chemoreception on the foraging behaviour of two species of sandpiper, *Calidris alba* (Pallas) and *Calidris alpina* (L.), *Netherlands Journal of Sea Research*, 17, 47–56.

Verner, J., and Willson, M. F., 1966, 'The influence of habitats on mating systems of North American passerine birds in the nesting cycle', *Ecology*, 47, 143–7.

Viguier, C., 1882, 'Le sens d l'orientation et ses organes chez les animaux et chez l'homme', *Revue philosophique de la France et de l'etranger*, 14, 1–36.

Villard, P., and Cuisin, J., 2004, 'How do woodpeckers extract grubs with their tongues? A study of the Guadeloupe woodpecker (*Melanerpes herminieri*) in the French Indies', *The Auk*, 121, 509–14.

Voss, H. U., Tabelow, K., Polzehl, J., Tchernichovski, O., Maul, K. K., Saldago-Commissariat, D., Ballon, D., and Helekar, S. A., 2007, 'Functional MRI of the zebra finch brain during song stimulation suggests a lateralized response topography', *PNAS*, 104, 10667–72.

Walls, G. L., 1942, *The Vertebrate Eye and its Adaptive Radiation*, Bloomingfield Hills, MI: Cranbrook Institute of Science.

Walsh, S. A., Barrett, P. M., Milner, A. C., Manley, G., and Witmer, L. M., 2009, 'Inner ear anatomy is a proxy for deducing auditory capability and behaviour in reptiles and birds', *Proc. R. Soc. Lond. B*, 276, 1355–60.

Watson, J. B., 1908, 'The behaviour of noddy and sooty terns', *Papers from the Tortugas Laboratory of the Carnegie Institution of Washington*, 2, 187–255.

Watson, J. B., and Lashley, K. S., 1915, 'A historical and experimental study of homing', *Papers from the Department of Marine Biology of the Carnegie Institute of Washington*, 7, 9–60.

Weir, A. A. S., Kenward, B., Chappell, J., and Kacelnick, A., 2004, 'Lateralization of tool use in New Caledonian crows (*Corvus moneduloides*)', *Proc. R. Soc. Lond. B*, 271, S344–S346.

Wenzel, B. M., 1965, 'Olfactory perception in birds', in *Proceedings of the Second International Symposium on Olfaction and Taste*, Wenner-Gren Foundation, New York, NY: Pergamon Press.

— 1968, 'The olfactory prowess of the kiwi', *Nature*, 220, 1133–4.

— 1971, 'Olfactory sense in the kiwi and other birds', *Annals of the New York Academy of Sciences*, 188, 183–93

— 2007, 'Avian olfaction: then and now', *Journal of Ornithology*, 148 (Suppl. 2), S191–S194.

Wheldon, P. J., and Rappole, J. H., 1997, 'A survey of birds odorous or unpalatable to humans: possible indications of chemical defense', *Journal of Chemical Ecology*, 23, 2609–33.

White, G., 1789, *The Natural History of Selborne*.

Whitfield, D. P., 1987, 'The social significance of plumage variability in wintering turnstones *Arenaria interpres*', *Animal Behaviour*, 36, 408–15.

Whitteridge, G., 1981, *Disputations Touching the Generation of Animals*, Oxford: Blackwell.

Wiklund, C., and Järvi, T., 1982, 'Survival of distasteful insects after being attacked by naive birds: a reappraisal of the theory of aposematic coloration evolving through individual selection', *Evolution*, 36, 998–1002.

Wild, J. M., 1990, 'Peripheral and central terminations of hypoglossal afferents innervating lingual tactile mechanoreceptor complexes in fringillidae', *The Journal of Comparative Neurology*, 298, 157–71.

Wilkinson, R., and Birkhead, T. R., 1995, 'Copulation behaviour in the vasa parrots *Coracopsis vasa* and *C. nigra*', *The Ibis*, 137, 117–19.

Wilson, A., and Ord, G., 1804–14, *American Ornithology*, Philadelphia, PA: Porter & Coates.

Wiltschko, R., and Wiltschko, W., 2003, 'Avian navigation: from historical to modern concepts', *Animal Behaviour*, 65, 257–72.

Wiltschko, W., and Wiltschko, R., 1991, 'Orientation in birds' magnetic orientation and celestial cues in migratory orientation', in *Orientation in Birds* (ed. Berthold, P.), pp. 16–37. Basel: Birkhauser Verlag.

— 2005, 'Magnetic orientation and magnetoreception in birds and other animals', *Journal of Comparative Physiology A*, 191, 675–93.

Winterbottom, M., Burke, T., and Birkhead, T. R., 2001, 'The phalloid organ, orgasm and sperm competition in a polygynandrous bird: the red-billed buffalo weaver (*Bubalornis niger*)', *Behavioural Ecology and Sociobiology*, 50, 474–82.

Wisby, W. J., and Hasler, A. D., 1954, 'Effect of occlusion on migrating silver salmon (*Oncorhynchus kisutch*)', *Journal of the Fisheries Research Board*, 11, 472–8.

Witt, M., Reutter, K., and Miller, I. J. M., 1994, 'Morphology of the peripheral taste system', in *Handbook of Olfaction and Gustation* (ed. Doty, R. L.), pp. 651–78, London: CRC.

Wood, C. A., 1917, *The Fundus Oculi of Birds Especially As Viewed by the Ophthalmoscope*, Chicago, IL: The Lakeside Press.

— 1931, *An Introduction to the Literature of Vertebrate Zoology*, London: Oxford University Press.

Wood, C. A., and Fyfe, F. M., 1943, *The Art of Falconry*, Stanford, CA: Stanford University Press.

Woodson, W. D., 1961, 'Upside down world', *Popular Mechanics*, January 1961, 114–15.

Young, L. J., and Wang, Z., 2004, 'The neurobiology of pair bonding', *Nature Neuroscience*, 7, 1048–54.

Zeki, S., 2007, 'The neurobiology of love', *FEBS Letters*, 281, 2575–9.

Glossary

Allopreening Preening the feathers of another bird; called 'allogrooming' in mammals.

Amplitude The loudness of a sound; measured as the amount of energy in a sound wave.

Anosmatic Loss of the sense of smell; smell-blind.

Anthropomorphism Attributing human characteristics to other animals.

Aposematic colouration A conspicuous colour pattern that warns of an animal's toxicity.

Attenuation Reduction of the intensity of sound over distance.

Audiogram Also known as an audibility curve. A graph showing frequency on the horizontal axis and hearing level (in decibels), from loud to quiet, on the vertical axis; used particularly to illustrate the softest sounds that can be heard.

Automaton A self-operating machine.

Basilar membrane The stiff membrane inside the cochlea of the inner ear that holds the sensory hairs (hair cells) involved in hearing.

Behavioural ecology The study of the behaviour within an ecological and evolutionary framework.

Brood parasite A bird (such as the European cuckoo) that parasitises the parental care of other bird species.

Brood patch An area of featherless skin on the abdomen of a bird through which heat is transmitted to incubate the egg(s). Birds may have one, two or three brood patches.

Cloacal protrusion   The cloacal region of the male vasa parrot inserted into the female during copulation to form a copulatory tie.

Cochlea   The elongated and often coiled (in mammals, but not birds) portion of the inner ear containing the sound-receptive cells.

Conchae   See 'nasal concha'.

Contour feathers   The outermost feathers covering the body.

Degradation of sound   Bird song (and other sounds) degrades over distance as a result of factors such as wind and vegetation; as a result, the further away one is from the source of a sound, the more muddled it sounds.

Distasteful insect   An insect that tastes unpleasant and/or is poisonous or has a painful sting.

Emlen funnel   Also known as an orientation cage; used to study migration behaviour in birds. Named after John T. and Steven T. Emlen (father and son, respectively), who invented it in the 1960s, the circular, funnel-shaped cage has an ink pad at the bottom and paper walls on which the bird leaves an inky trace with its feet, indicating the direction and intensity of its migratory behaviour.

Endocrine system   The system of glands that secrete hormones (chemical messengers) into the bloodstream.

Eustachian tube   The tube that connects the throat to the middle ear.

Filoplume   Hair-like feathers; one of several feather types.

Fovea   A pit in the retina at the back of the eye; the region of maximum visual acuity.

Fundus oculi   The concave interior of the back of the eye.

Geolocator   A miniature archival light recorder – a light-level logger – used for tracking animal movements. It works by recording the timing of dawn and dusk, from which latitude and longitude can be estimated.

Grandry corpuscles   Touch receptors in the beak and tongue of birds.

Herbst corpuscles   Touch receptors in the skin and beak of birds, typically larger than Grandry corpuscles.

Hypothalamus A gland within the brain that controls the digestive and reproductive systems and regulates many behaviours, such as feeding.

Imprinting A form of learning that usually occurs within a specific time window (the sensitive period) early in an individual's life. Filial imprinting is when offspring learn who their parents are; sexual imprinting occurs when individuals learn characteristics that they will later use when choosing a sexual partner, usually learned by viewing their mother and father.

Lateralisation The tendency to use one eye or hand more than the other.

Macula The region of the retina in the eye that contains the fovea.

Nasal concha A thin, scroll-like bone in the beak of birds, covered by a thin layer of tissue (the nasal epithelium) in which the olfactory receptors are located. Singular: concha; plural: conchae.

Neurohormone A hormone released from specialised nerve cells (neurosecretory cells) into the blood, rather than being released from endocrine glands into the blood. Oxytocin is an example of a neurohormone that is produced in the brain.

Nictitating membrane A transparent or translucent third eyelid in birds and other vertebrates; rare in mammals.

Passerine Also known as perching birds, or, less precisely, songbirds. Passerines comprise more than half of all birds (cf. non-passerines); they include the true songbirds and suboscines, such as the New World flycatchers.

Pecten A structure, often pleated or comb-like, within the posterior chamber of the eyes of birds.

Phalloid organ A penis-like structure in two buffalo weaver bird species, larger in the male than the female, lying on the anterior edge of the cloaca.

Photosensitive cells Light receptors – rods and cones; specialised cells in the retina of the eye.

Phylogenetic effect If all members of a taxon (such as a genus or family) exhibit the same feature (such as clutch size or number of tail feathers), it is said to be a phylogenetic effect, meaning that all members of the taxonomic unit possess it because they share a common ancestor.

Polygyny A type of mating system in which a male has more than one female partner: a form of polygamy. Other mating systems include monogamy, where one male and one female pair together, and polyandry, where a female has more than one male partner.

Rictal bristle Stiff, hair-like feathers located near the mouth (rictus).

Sonogram A graphical image of sound produced by a sonograph or sound spectrograph machine, showing frequency (or pitch) on the vertical axis and duration on the horizontal axis; used to analyse birdsong.

Visual acuity Refers to the sharpness of vision or spatial resolution of an image.

Visual sensitivity The ability to discriminate objects at low light levels.

Index

A NOTE ON THE AUTHOR

Tim Birkhead is a professor at the University of Sheffield where he teaches animal behaviour and the history of science. He is a Fellow of the Royal Society of London and his research has taken him all over the world in the quest to understand the lives of birds. He has written for the *Independent, New Scientist* and *BBC Wildlife*. Among his other books are *Promiscuity, Great Auk Islands, The Cambridge Encyclopaedia of Ornithology* which won the McColvin medal, *The Red Canary* which won the Consul Cremer Prize and *The Wisdom of Birds* which was voted 'Best Bird Book of Year' (2009) by the British Trust for Ornithology and British Birds. He is married with three children and lives in Sheffield.